66 Topics in Current Chemistry
Fortschritte der chemischen Forschung

Triplet States III

Springer-Verlag
Berlin Heidelberg GmbH 1976

This series presents critical reviews of the present position and future trends in modern chemical research. It is addressed to all research and industrial chemists who wish to keep abreast of advances in their subject.

As a rule, contributions are specially commissioned. The editors and publishers will, however, always be pleased to receive suggestions and supplementary information. Papers are accepted for "Topics in Current Chemistry" in either German or English.

ISBN 978-3-662-15982-8 ISBN 978-3-540-38192-1 (eBook)
DOI 10.1007/978-3-540-38192-1

Library of Congress Cataloging in Publication Data (Revised) Main entry under title: Triplet states. (Topics in current chemistry; 54-55, 66) Includes bibliographies and indexes. CONTENTS: 1. Devaquet, A. Quantum-mechanical calculations of the potential energy surfaces of triplet states. Ipaktschi, J., Dauben, W.G., and Lodder, G. Photochemistry of B. γ-unsaturated ketones. Maki, A. H. and Zuclich, J. A. Protein triplets states. [etc.] 1. Excited state chemistry-Addresses, essays, lectures. 2. Triplet state-Addresses, essays, lectures. I. Series: Topics in current chemistry; 54 [etc.] QDl. F 58 vol. 54-55 [QD461.5] 540'.8s [541'.24] 75-1466

Originally published by Springer-Verlag Berlin Heidelberg New York in 1976.

Softcover reprint of the hardcover 1st edition 1976

Contents

Springer-Verlag

Postfach 105280 · D-6900 Heidelberg 1
Telephone (06221) 487-1 · Telex 04-61723
Heidelberger Platz 3 · D-1000 Berlin 33
Telephone (030) 822001 · Telex 01-83319

Springer-Verlag New York Inc.

175, Fifth Avenue · New York, NY 10010
Telephone 673-2660

Chemistry of Excited Triplet Organic Carbonyl Compounds

Professor Peter J. Wagner

Chemistry Department, Michigan State University, East Lansing, Michigan, U. S. A.

Contents

I. Introduction

Anyone who has ever taught organic chemistry realizes how important and varied are the reactions of carbonyl compounds. It is not surprising that studies of carbonyl compounds, particularly ketones, have played a very large role in the development of synthetic and mechanistic photochemistry. Not only do ketones and aldehydes absorb light in the convenient near ultraviolet, they can be prepared with seemingly infinite structural variations and they undergo a wide variety of synthetically useful photoreactions. The wide variety of *measurable* chemical and physical reactions undergone by excited ketones has allowed extensive correlations of structure with reactivity, such that the photochemistry of the carbonyl group is understood far better than that of any other functional group.

The photochemistry of acid derivatives is not well understood. Since these compounds absorb in the far ultraviolet and rarely react with great quantum efficiency, they have not been studied at all systematically. Consequently, this review will emphasize aldehyde and especially ketone photochemistry. So much is known about carbonyl photochemistry and so much is added each year (*e. g.*, over 200 pages in the Chemical Society's 1973 survey of photochemistry [1]), that this review had to be limited to but a few aspects of carbonyl photochemistry. Although the main kinds of carbonyl photoreactions have been adequately reviewed often [2], several important generalizations have resulted from the past few years' work. This review will attempt to summarize these new generalizations which may not yet be familiar to the entire chemical community. Several unresolved problems will also be outlined.

The dynamics of photochemical reactions are generally considered from a statistical viewpoint. Once a molecule is electronically excited by absorption of a photon, it can undergo several competing physical decay reactions. The most important of these are:

internal conversion to lower excited states;

intersystem crossing between singlet and triplet manifolds;

emmision of light (fluorescence or phosphorescence);

radiationless decay to the ground state; and

energy transfer to another chromophore.

Any chemical reaction of a particular excited state must compete with these physical decay modes. For a photoreaction to occur at all efficiently, some excited state of the reactant must undergo an irreversible chemical reaction at a rate comparably fast to those of the physical decay reactions. Knowledge of rate constants is much more important in photochemistry than it is in ground state chemistry because excited states, unlike ground states, do not sit around patiently waiting for someone or something to make them react. A general understanding of the photoreactions of the carbonyl group demands a good understanding of the nature of the different possible carbonyl excited states and how the rates of the various chemical and physical reactions of these excited states vary with structure.

The quantum yield of a photoreaction represents the probability that light absorption will produce that reaction. Not only does the quantum yield measure the efficiency of a photoreaction, it is the only kinetic parameter which can be measured under steady state conditions, since photoreactions generally follow zero order kinetics. For a triplet reaction, the quantum yield can be dissected into three separate probabilities, any or all of which can be less than unity [3].

$$\Phi_R = \phi_{isc} \cdot \phi_R \cdot P_P \tag{1}$$

$$\Phi_R = k_{isc}\, \tau_S \cdot k_r\, \tau_T \cdot P_P \tag{2}$$

The intersystem crossing yield ϕ_{isc} is the fraction of singlets which cross over to the lower energy triplet instead of reacting themselves, fluorescing, or otherwise decaying. This quantity can usually be measured independently by using the compound to photosensitize some well known triplet reaction of another compound, such as the phosphorescence of biacetyl [4] or the *cis-trans* isomerization of an olefin [5]. The only requirement is that the triplet of the sensitizer be long-lived enough to transfer energy to practicable concentrations of the acceptor. The rate constant k_{isc} for intersystem crossing is large ($>10^8$ sec^{-1}) for all ketones, so that they often form triplets in high yield. For this reason, many phenyl ketones are among the most popular photosensitizing agents. With aliphatic ketones, fluorescence, intersystem crossing, and chemical reaction are often competitive reactions of the lowest excited singlet so that study of these compounds has provided a wealth of information about differences in reactivity between singlets and triplets.

The probability ϕ_R that an excited state will react is defined according to its rate parameters simply as $k_r\tau$, where k_r is the rate constant for the reaction and τ is the kinetic lifetime of the excited state (τ_T for the triplet), determined by the rates of *all* reactions undergone by that state. The unimolecular physical reactions of excited triplets are spin-forbidden and therefore relatively slow, so that chemical reactions are often highly efficient ($k_r\tau_T \approx 1$).

Despite high probabilities for triplet formation and for triplet reaction, many overall photoreactions proceed in low quantum efficiency. In the absence of competing chemical reactions, the factor most often responsible for low quantum yields is the revertibility [3] of primary triplet reactions. Metastable intermediates such as radicals, biradicals, and charge transfer complexes (either excited or ground state) are the usual photoproducts from excited triplet reactions. These intermediates generally can revert to ground state reactant, thus providing a chemical path for "radiationless decay", as well as proceed to stable products. Hence, the factor P_p is necessary to describe the probability that the intermediate will form product.

There is one physical reaction of triplets which is not slow. Bimolecular exothermic energy transfer proceeds at rates which approach those of diffusion control [6]. It is thus possible to selectively quench out the triplet photoreactions of a given compound by adding a compound which has a lower triplet excitation energy but which does not itself absorb significantly at the wavelength being used. Conjugated dienes have been most often employed to quench aliphatic ketones. 1,3-Pentadiene is very useful because it is a relatively poor quencher of excited singlets [7] and it undergoes easily measurable geometric isomerization [5]. Its only serious drawback is that it is a good radical scavenger. 2,5-dimethyl-2,4-hexadiene does not suffer from this drawback but is a rather efficient singlet quencher [7]. With phenyl ketones polynuclear aromatics such as naphthalene can be employed. Actually, aromatics such as biphenyl and naphthalene can be used to quench even aliphatic ketone triplet state photochemistry. Even though these aromatics have much higher extinction coefficients than the ketones at all wavelengths, they transfer their singlet excitation to the ketones [8]. With the yellow α-diketones, aromatics such as anthracene and pyrene are good triplet quenchers. Molecular oxygen quenches most excited triplets; even the small concentrations normally present in air-equilibrated solutions are sufficient to quench triplets which do not react within one μsec. Lamola's review on energy transfer [6] tabulates the triplet energies of many readily available potential quenchers.

The efficiency of quenching is proportional to the rate constant for energy transfer, k_q, and the lifetime τ of the excited triplet being quenched. Since k_q values are so close to being diffusion controlled, quenching studies provide a simple method for measuring triplet lifetimes.

There are two distinct types of electronic excitation which occur in carbonyl compounds and which produce excited states with quite different physical and chemical properties. In aliphatic ketones and aldehydes, the weak absorption with λ_{max} at 280—300 nm represents an $n \rightarrow \pi^*$ transition, in which a nonbonding electron on oxygen is promoted into a π^* orbital. The resulting excited state resembles a diradical both in its physical and chemical properties. In particular, the chemical reactions of triplet n,π^* states parallel those of alkoxy radicals [2,9], almost as if the electron in the antibonding orbital does not affect the electron left on oxygen.

It is perhaps valuable to compare electronic excitation with one-electron redox processes. For example, an $n \rightarrow \pi^*$ excitation is equivalent to oxidizing the n-orbital and reducing the π-system. The predominant valence bond structure of the

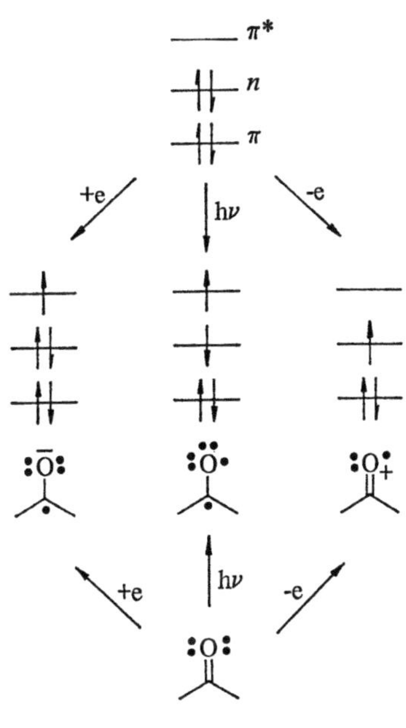

radical anion has an odd electron on the carbon and a negative charge on oxygen; that of the radical cation has the odd electron and the positive charge both on oxygen; combining the two gives the diradical. Inasmuch as an electron in the π^* orbital results in a partial negative charge on carbon, the n,π^* state has another

resonance form which emphasizes that the oxygen atom as well as its n orbital is electron-deficient, while the carbon is somewhat electron-rich. It has been noted

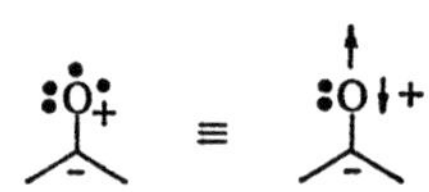

by several authors that n,π^* singlet carbonyls demonstrate marked nucleophilicity at carbon, whereas n,π^* triplets behave more like diradicals. This difference is probably due to the necessity for electrons with parallel spin to get as far apart as possible, such that the zwitterionic resonance form is destabilized in the triplet but not in the singlet.

In conjugated ketones, the $n \rightarrow \pi^*$ transition occurs at lower energy, with λ_{max} occuring at 325 nm for typical phenyl ketones. In terms of triplet energies, those of aliphatic ketones are usually taken to lie in the 76—80 kcal/mole range. A more precise figure is not obtainable because the phosphorescence of aliphatic ketones shows no vibrational structure so that a 0—0 band energy is not assignable. Aldehydes presumably have lower n,π^* triplet energies, since that of formaldehyde is 72 kcal/mole [10)]. The n,π^* E_T values of phenyl alkyl ketones lie around 72 kcal [11)]. while that of benzophenone lies at 68 kcal [12)]. The n,π^* triplets of conjugated enones also lie 70—72 kcal above their ground states [13)].

Although the $\pi \rightarrow \pi^*$ transitions of aliphatic ketones occur at very high energy, those of conjugated ketones lie much lower. In particular, phenyl ketones have π,π^* triplets which lie less than 75 kcal above the ground state [14)], while enones have π,π^* triplets which lie within a few kcal either way of their n,π^* triplets [13)]. Substituents on the benzene ring of phenyl ketones or on the double bond of enones independently stabilize or destabilize the n,π^* and π,π^* transitions, so that such substituents usually determine which of the two triplets is lower in energy. The polarity of the solvent also affects relative triplet energy levels, with π,π^* triplets being stabilized in hydrogen-bonding and high dielectric solvents [14)]. When an acyl group is attached to any aromatic ring bigger than benzene, the lowest triplet of the compound is basically a perturbed π,π^* triplet of the aromatic.

The excited state behavior of a compound usually depends largely on the nature of its lowest triplet. In compounds where the two triplets are energetically proximate, they can equilibrate thermally before decaying and they may also mix with each other. How much these two phenomena affect chemical reactivity is still the subject of active research and speculation. The latest thinking is that mixing is not significant unless the two "pure" triplets would be within 0.5 kcal of being isoenergetic [15]. Consequently, it is usually fairly safe to think of these triplets in terms of a simple m.o. model. An orbital diagram and commonly accepted valence structures are presented below for the *n*,*π** and *π*,*π** states of a simple enone. The *π*,*π** triplet is more nucleophilic than electron-deficient at oxygen and may have diradical character. Inasmuch as the chemical reactions of such states reflect the electronic distribution of the reactant, quite different chemical reactivity is expected and observed for *n*,*π** and *π*,*π** triplets.

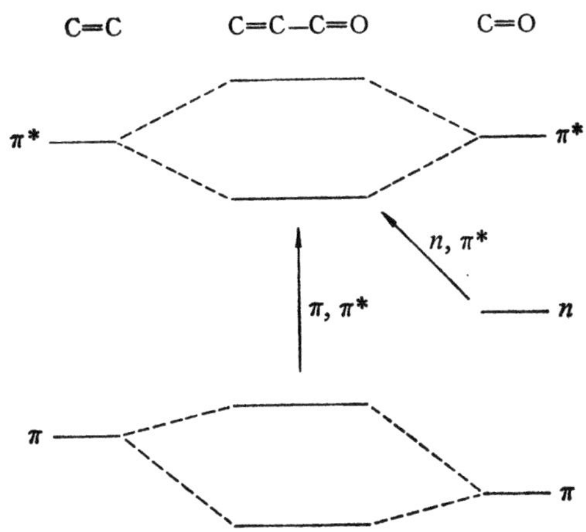

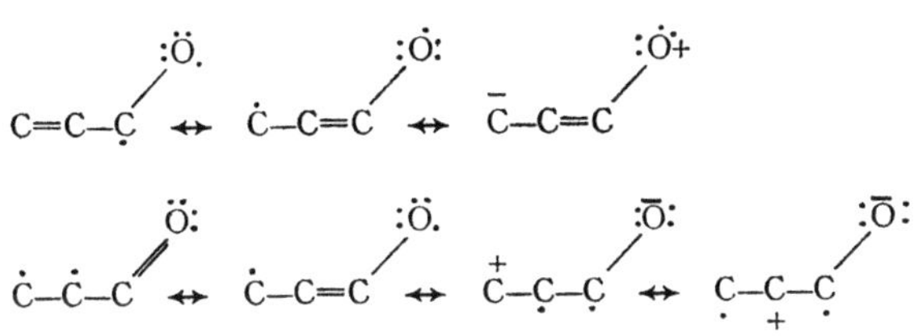

n,*π**

π,*π**

II. Primary Photoreactions

The photochemistry of alkanones has been reviewed recently [2c]. It would appear that rate constants for the physical decay processes of both singlets and triplets vary only slightly among a wide collection of structurally varied acyclic, cyclic, and polycyclic alkanones. Consequently, relative triplet yields are determined primarily by relative rates of excited singlet chemical reactions. As in most triplet reactions, overall quantum yields are usually determined primarily by the partitioning of metastable primary photoproducts.

All of the photoreactions of aliphatic carbonyl compounds result from just four primary reactions of *n*,*π** states. More complicated carbonyl compounds can undergo various rearrangement reactions and reactions characteristic of *π*,*π** states but also undergo the four basic carbonyl reactions. A brief summary of these reaction types is presented in this section; the remainder of this review is devoted to a critical summary of how structure affects the rates and efficiencies of these basic reactions. Since each of the primary photoreactions produces another reaction intermediate, the chemistry of both the excited states and the intermediates must be understood if quantum efficiencies are to be predictable.

A. α-Cleavage is the common unimolecular, intrachromophoric radical-producing reaction of the excited carbonyl group and is closely analogous to the β-cleavage of alkoxy radicals [16]. Acyclic compounds yield a pair of radicals

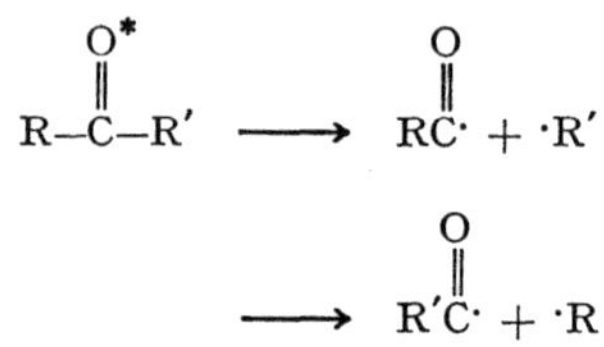

which can either react in the initial solvent cage or diffuse apart and initiate typical radical reactions. Cyclic compounds yield diradicals which can either recouple or disproportionate to yield stable products.

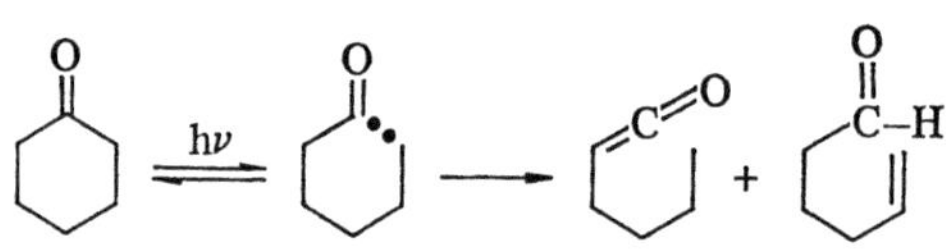

B. Hydrogen Atom Abstraction is the other alkoxy-radical-like reaction of triplet n,π^* carbonyls. The substrate can be another molecule, in which case a pair of radicals is formed; or an accessible hydrogen in the same molecule, in which case a diradical is formed.

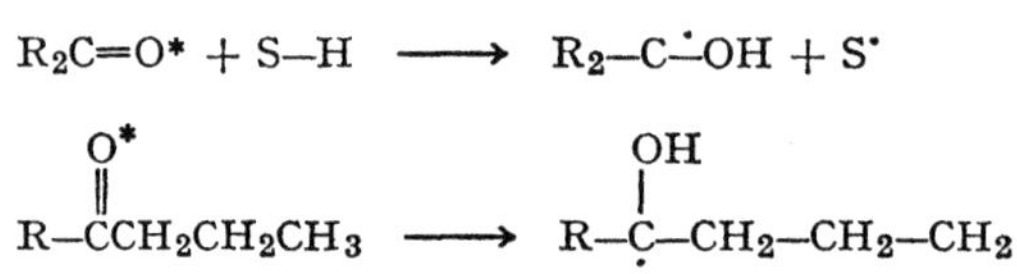

C. Charge Transfer Complexation. The one-electron oxidation and reduction potentials of excited triplets are lower than those of the ground state by the triplet excitation energy E_T. Therefore, excited triplets (and all excited states) are very prone to get involved in charge transfer and electron transfer processes. Charge transfer complexation seems to account for most bimolecular chemical reactions of ketone triplets which are not initiated by hydrogen abstraction. There is as yet no hard evidence whether the CT complexes so formed are still electronically excited (exciplexes) or whether they represent some high energy ground state configuration, but exciplexes certainly are likely intermediates. Depending on the nature of the substrate, the complex can proceed to products by various chemical paths; but reverse electron transfer always operates to quench a significant fraction of the complexes.

D. β-Cleavage. Ketones substituted with heteroatoms on an α-carbon, including α-halo, α-tosyl and α-aminoketones, undergo photoinduced loss of that substituent either as a radical or as an anion [17].

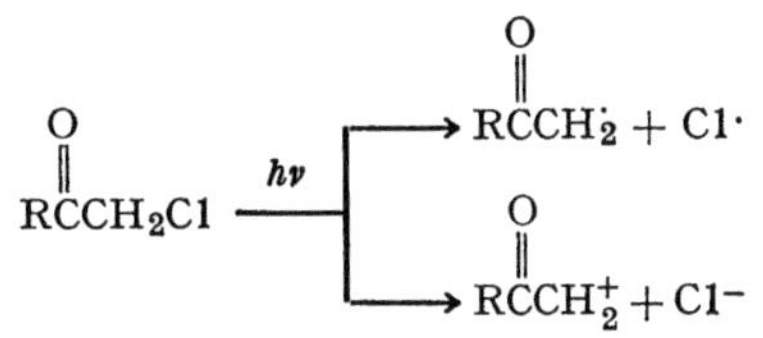

III. α-Cleavage

A. Acyclic Aliphatic Ketones

The photolysis of ketones into radicals is called Norrish type I cleavage. The efficiency of this cleavage depends strongly on the ketone's structure and on the medium.

The simplest ketone, acetone, has remained one of the hardest to understand. It photofragments efficiently in the gas phase with unit quantum yield above 100 °C [18]. In inert solvents, however, it fragments so inefficiently [19] as to require CIDNP techniques to be detected [20]; even then, only in water is CIDNP readily detected for the products expected from cleavage. (In solvents with labile C—H bonds, acetone is photoreduced [21]).

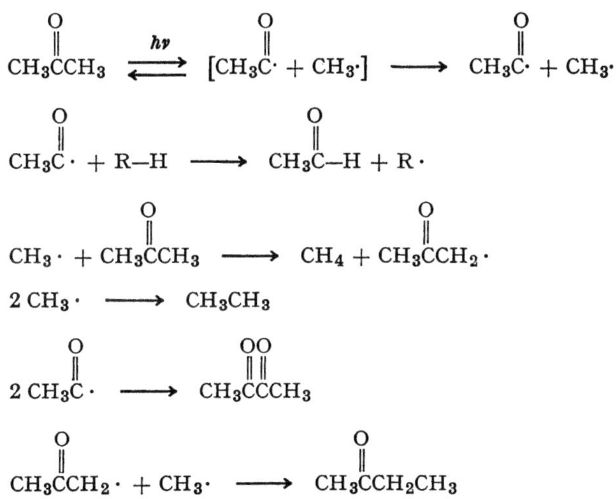

Unlike acetone, diethyl ketone cleaves well even in cumene [22]. 2-Pentanone triplet undergoes considerable α-cleavage in competition with rapid intramolecular hydrogen abstraction, as judged by the buildup of an efficient triplet quencher [23]. Biacetyl is the only likely candidate for that quencher and is the major product of α-cleavage of methyl ketones at temperatures low enough that decarbonylation of the acetyl radical is slow. Isopropyl, *tert*-butyl, and benzyl ketones all cleave quite efficiently and various reports have appeared on the CIDNP spectra of products derived from the radicals [24].

The rate constant for triplet α-cleavage depends strongly on the stability of the radical produced, as judged by both intramolecular competition and by measured triplet lifetimes. An order of magnitude comparison of rate constants can be estimated for cleavage of progressively more stable radicals R· as follows: CH_3, 10^3—10^4 sec^{-1}; CH_3CH_2, $\sim 10^6$—10^7 sec^{-1}; $(CH_3)_2CH$, $\sim 10^8$ sec^{-1}; $C(CH_3)_3$; 2×10^9 sec^{-1}; $PhCH_2$, 10^{10} sec^{-1}. The last two are measured [25,26]; the second would explain the above-mentioned behavior of 2- and 3-pentanone.

$$CH_3\overset{\overset{O^*}{\|}}{C}R \longrightarrow CH_3\overset{\overset{O}{\|}}{C}\cdot + R\cdot$$

The values for methyl and isopropyl have not been even indirectly measured. They were chosen by analogy to the known relative rates of β-cleavage of alkoxy radicals [27]. The number so chosen for methyl explains the inefficient cleavage of acetone in solution. Although this rate is large enough to compete with quenching by unreactive solvents, it is not large enough to compete with quenching by product biacetyl. By the time 10^{-6}M biacetyl is produced (0.001% conversion for 0.1 M acetone), the reaction is half-quenched by product, since k_q is on the order of $10^{10}M^{-1}$ sec^{-1} in most common solvents [6]. Although quenching by product has long been recognized as an *a priori* possibility in all photoreactions, it is a serious complication surprisingly rarely.

It is noteworthy that the excited singlets of *t*-butyl ketones undergo α-cleavage only 1/100 as fast as their triplets [25], despite their 5 kcal greater energy. This fact has been a real puzzle for some time and remains so today. Turro suggested that the triplet ketone can correlate with a repulsive triplet of the radical pair [28,29]. Salem has recently applied simple symmetry considerations to this photocleavage reaction [29] and concluded that the singlet radical pair with the same symmetry as the n,π^* ketone singlet is a high energy state, whereas the n,π^* triplet can indeed pass smoothly to the lowest energy triplet radical pair either by mixing with an upper triplet or by losing all symmetry by cleaving in a non-coplanar fashion. Other calculations [30] and some experiments [31] suggest that the latter is the case.

Decarbonylation regularly accompanies α-cleavage reactions at higher temperatures in the gas phase but occurs at room temperature in solution only if the acyl radical can form a very stable alkyl radical (tert-alkyl, benzyl). For example, di-*tert*-butyl ketone photofragments with a total quantum yield of 0.7 in solution [25]. Cage recombination invariably cuts the efficiency of type I cleavage in solution. In terms of the general kinetic scheme, P_p is less than unity.

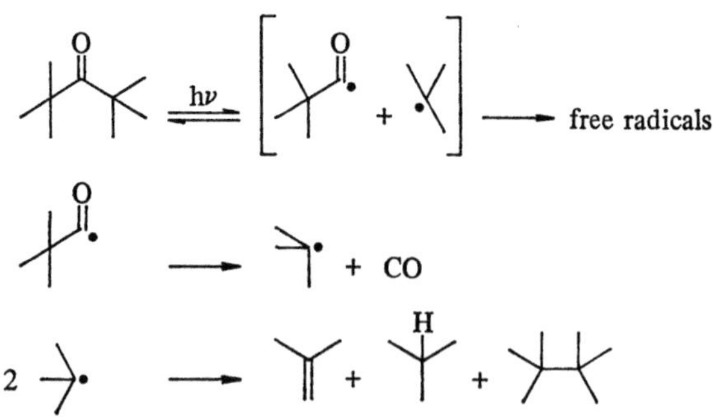

If decarbonylation of the acyl radical is slow, aldehydes and diketones are formed.

B. Aldehydes

Photoreactions of aldehydes have a strong tendency to become messy if radicals are produced, because of the high reactivity of aldehydic hydrogens in abstraction reactions. β,γ-unsaturated aldehydes undergo a highly stereospecific decarbonylation, which probably takes place from the excited singlet [32].

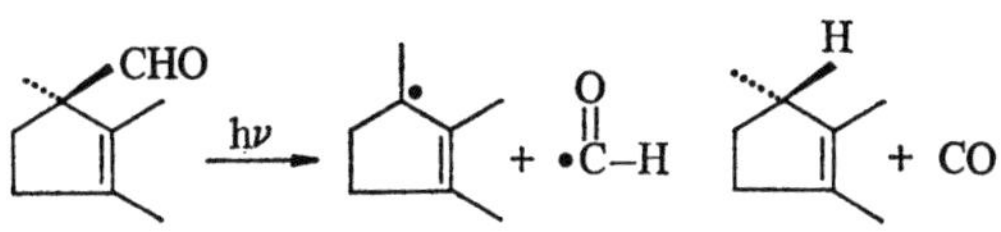

C. Cyclic Ketones

α-Cleavage of a cyclic ketone necessarily produces a diradical. It is now well accepted that the photorearrangement of cycloalkanones to ω-unsaturated aldehydes and to ketenes proceeds via such diradicals. It has not always been adequately appreciated that a diradical mechanism had to be a possibility for this rearrangement given that excited ketones cleave into radicals. The only way the rearrangement could be primarily or exclusively a concerted process would be if that process were much faster than the homolytic cleavage.

The evidence for diradicals in cycloalkanone photochemistry is widespread. Proton [33], fluorine [34], and ^{13}C-CIDNP [35] spectra have been recorded for products and reactant ketone. This evidence indicates that a diradical is at least partially involved in the rearrangement as well as in the well-known photoepimerization of certain keto-steroids [36]. Such loss of stereochemistry is usually associated

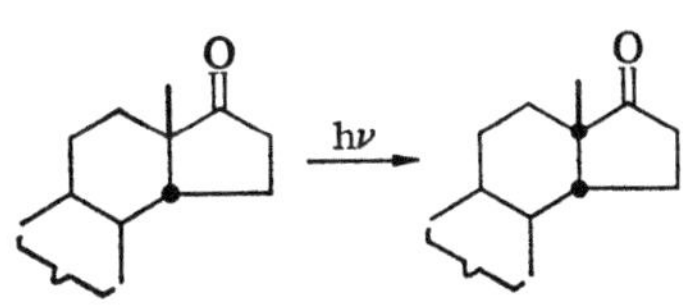

with the presence of an sp^2-hybridized intermediate. Here a diradical is the obvious choice. The identical product distribution from both *cis* and *trans*-2,3-dimethylcyclohexanone is a more subtle indication that there is an intermediate involved in the rearrangement which loses its configurational integrity at carbon 2 [37].

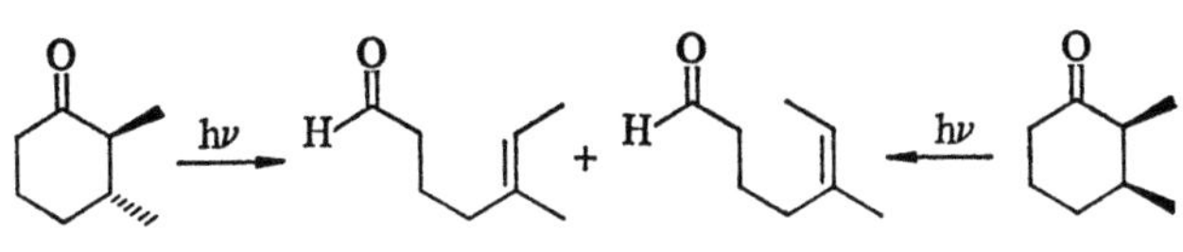

The kinetics of these rearrangements demand an intermediate. Triplet lifetimes depend strongly on α-substitution [38,39] in line with what would be expected for relative stabilities of the alkyl-half of the diradical. 3- and 4-Substituents have only small effects on rates of triplet decay. Most important, there is no correlation between quantum yield and triplet state reactivity [38], symptomatic behavior for photoreactions in which an intermediate reverts to starting material.

Relative rates of α-cleavage to yield primary: secondary: tertiary: benzylic alkyl radical sites are 1:30:300:1000, just as with the α-cleavage of acyclic ketone triplets into radicals. For the cyclohexanones, the absolute rates are very similar to these for α-cleavage of acyclic ketones, as befits a similar mechanism. The rate for 2-phenylcyclohexanone is considerably larger than first reported [38], more in line with the behavior of dibenzyl ketone [26]. Cyclopentanones cleave ten times faster than the corresponding cyclohexanones, presumably manifesting the release of ring strain. It is noteworthy that excited singlet lifetimes *increase* with α-substitution [39]. The rearrangements of the cyclic ketones are therefore predominantly triplet reactions, although the fact that part of the cleavage of α,α-dialkylcyclohexanones is inquenchable [40] indicates some singlet α-cleavage. In this respect, it is worth mention that application of Kaptein's rule to the CIDNP spectra of Norrish type I cleavage reactions [24] suggests the intermediacy of triplets. Since this conclusion is derived only from the sign of polarization, it is impossible to rule out partial (<50%) singlet participation.

Most of the remaining uncertainties in cycloalkanone photochemistry are concerned with the behavior of the diradical intermediate. Its major reactions include recoupling and disproportionation in two senses to yield ketene or unsaturated aldehyde. A few strained cyclopentanones also partially recyclize to oxycarbenes [41].

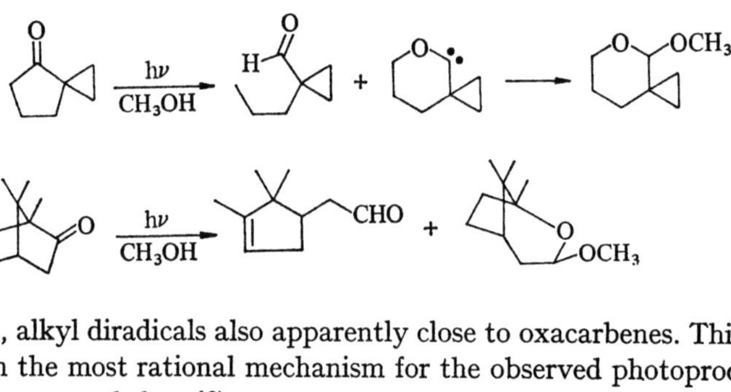

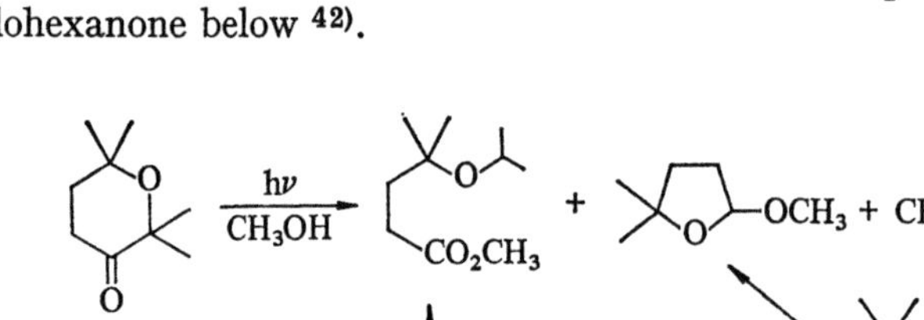

1,4-Acyl, alkyl diradicals also apparently close to oxacarbenes. This statement derives from the most rational mechanism for the observed photoproducts of the 3-oxacyclohexanone below [42].

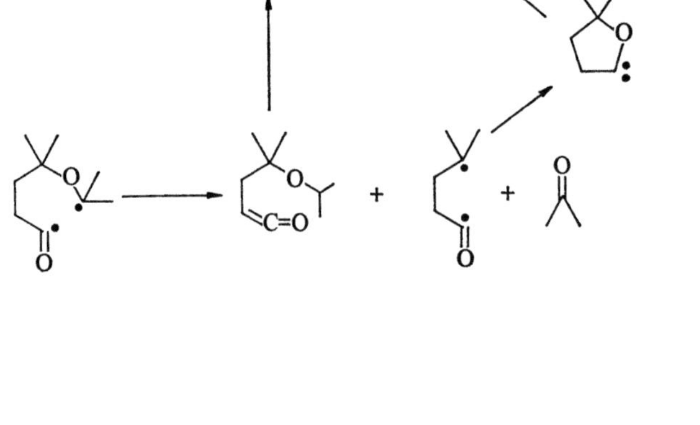

It is very difficult to predict how a given diradical will partition itself among its possible competing reactions, since any structural modification affects each competing reaction independently. 3- and 4-Alkyl groups apparently slow aldehyde formation in favor of ketene formation, whereas 2-substituents enhance aldehyde formation [38,39]. Rate constants for radical-radical reactions are little affected by alkyl substitution away from the trivalent carbon. The large changes in P_p values produced by 3- and 4-alkyl groups on cyclohexanones suggest that, in the intramolecular reactions of diradicals, rates of conformational change become at least partially product-determining. Each separate diradical reaction can occur from only a few of the many conformations accessible to the diradical. Heavy substitution about a given C—C bond could well slow rotation about that bond to the point that the diradical, which is born in a conformation similar to that of the parent ketone, cannot achieve the conformation required for disproportionation rapidly enough to compete with recoupling. It is difficult to think of any other explanation of the remote substituent effects. It must be emphasized that this interpretation necessitates that rates of bond rotations be comparable to those of coupling of *triplet*-derived diradicals. This problem will be discussed again later.

Cyclobutanones are unusual in that triplet sensitization experiments seem to indicate that α-cleavage is not efficient from the triplet [43,44]. Given the rapid cleavage of triplet cyclopentanones, it is difficult to explain why added ring strain does not enhance the rate of cleavage even more. Direct irradiation of cyclobutanone derivatives results in fragmentation and oxacarbene formation [45]. Both processes could easily arise from a 1,4-diradical, but extensive arguments have been proposed that a concerted reaction might be involved in the carbene formation [46]. These direct photoreactions probably involve excited singlets.

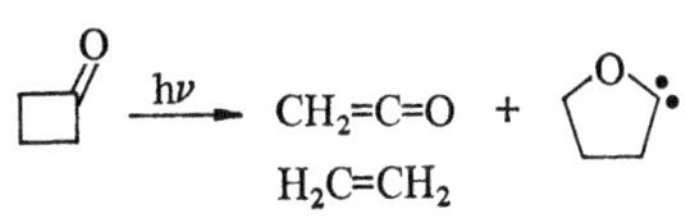

D. Phenyl Ketones

The α-cleavage reactions of aryl ketones have been studied carefully only recently. The enhanced rate of intersystem crossing in phenyl ketones usually guarantees that most reactions proceed from triplet levels. Several workers independently studied the type I cleavage of *tert*-alkyl phenyl ketones and found that the rate constants are only 1/100 as large as for the triplets of aliphatic ketones [47]. The most reasonable explanation for this difference is that the lower triplet energies of phenyl ketones makes the cleavage less exothermic. However, Heine and Lewis recently have done the first systematic studies on phenyl benzyl ketones. Phenyl benzyl ketone itself cleaves only 1/10 as rapidly as triplet pivalophenone, $k \sim 2 \times 10^6$ sec^{-1}, [48] despite the fact that loss of a benzyl radical should be some 6 kcal more exothermic than loss of a *tert*-butyl radical. The benzyl ketone triplet is so unreactive that it phosphoresces appreciably at room temperature. Although it has been suggested that steric effects rather than radical stability are of predominant importance in determining the rate constants for these sorts of radical cleavage reactions [48], it is more likely that the formation of benzyl radical is slowed by significant entropy loss in the transition state for cleavage [49].

Careful study of (S)-(+)-2-phenylpropiophenone reveals that approximately half of the radical pairs recombine before diffusing out of the initial solvent cage [50]. This conclusion follows from the 44% quantum yield of scavengable benzoyl radicals and the 33% quantum yield for racemization. Alkyl thiols are excellent radical scavengers in carbonyl photochemistry because they quench triplet ketones fairly slowly [51]. Lewis has shown that concentrations of thiol above 0.03 M generally trap all *free* benzoyl radicals as benzaldehyde [50]. Of course, the minimum concentration for complete scavenging depends on conversion.

The extent of cage recombination is of very practical importance, since benzoin ethers are among the most common initiators in several commercial photoinitiated polymerizations [52]. The excited states of these benzoin ethers are so reactive that they undergo no measurable bimolecular reactions.

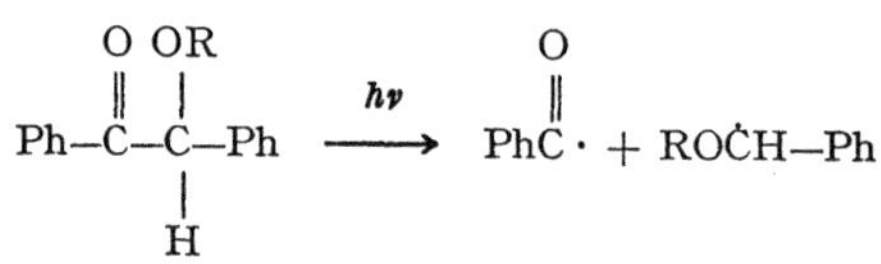

Only one study has been reported of the effect of electronic configuration on α-cleavage. Lewis reports that triplet p-methoxypivalophenone cleaves only 1/200 as rapidly as triplet pivalophenone itself, while p-phenylpivalophenone is quite stable to light [53]. The two *para*-susbtituted ketones have π,π^* lowest triplets. Their relative reactivities in α-cleavage parallel those in the much more extensively studied hydrogen abstraction reactions. This comparison indicates that n,π^* excitation is required for α-cleavage and furthers the generality of the simple alkoxy-radical like picture of n,π^* excited ketones.

Sophisticated theoretical pictures of carbonyl compounds all point out that the n-orbital and the C—C σ orbitals are mixed. Therefore, an n,π^* excitation intrinsically weakens the carbonyl-α-carbon bond and predisposes that bond to rupture. However, the actual cleavage process is certainly distinct from the electronic transition and apparently follows potential surfaces analogous to those followed by alkoxy radicals in their β-cleavage reactions. Mixing of the n-orbital with the C—C σ orbital is presumably very similar in the *transition states* for both kinds of cleavage and involves three electrons. The π^* electron in the excited state reaction is in an orbital orthogonal to the reaction coordinate and is not apparently involved in the reaction.

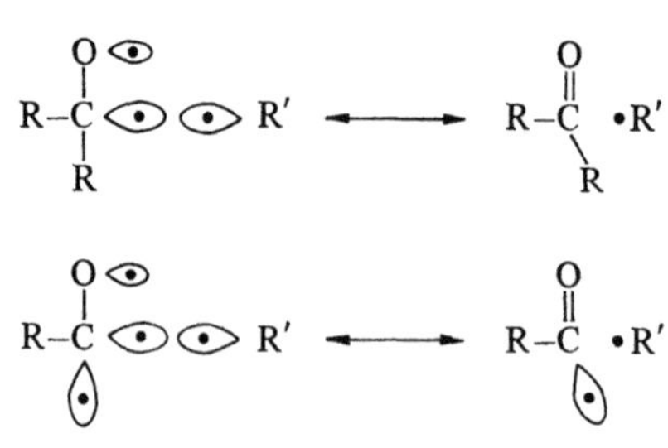

IV. Hydrogen Abstraction

Hydrogen abstraction by triplet carbonyl compounds has been the most widely studied excited state reaction in terms of structural variations in reactants. Consequently, the most detailed structure-reactivity relationships in photochemistry have been developed for hydrogen abstraction. These correlations derive from studies of both bimolecular reaction and intramolecular reactions. The effects of C—H bond strength and the inductive and steric effects of substituents have been analyzed. The only really quantitative comparisons between singlets and triplets and between n,π^* and π,π^* states have been provided by studies of photoinduced hydrogen abstractions.

A. Comparisons of *n*, *π** and *π*, *π** Triplets

Phenyl ketones invariably have an n,π^* and a π,π^* triplet of comparable energy. Because the $n \rightarrow \pi^*$ transition causes electron density to move away from oxygen and the $\pi \rightarrow \pi^*$ transition causes electron density to move from the benzene ring toward the carbonyl, ring substituents tend to have opposite inductive effects on n,π^* and π,π^* transition energies. Electron-donating substituents stabilize the π,π^* triplet and destabilize the n,π^* triplet. Polar solvents do the same. In the case of phenyl alkyl ketones, even one methyl group is enough to invert the triplet levels; with benzophenone, only strongly electron-donating groups such as amino make the lowest triplet π,π^* in character. Groups which are inductively electron-withdrawing (such as CF_3, m-CN) increase the separation between triplets by stabilizing the n,π^* state. Electron-withdrawing groups which can conjugate with the carbonyl in the *para*-position, such as halogens and cyano, stabilize both n,π^* and π,π^* transition.

Acylbiphenyls and acylnaphthalenes have lowest π,π^* triplest with the nearest n,π^* triplet being >6 kcal higher. The greatly reduced photoreactivity of naphthyl ketones relative to benzophenone [54] was the earliest evidence that π,π^* triplets are not good hydrogen atom abstractors. Yang was the first to point out that methyl- and methoxy-substituted acetophenones, with π,π^* lowest triplets, are efficiently photoreduced [55]. Later quantitative work established that phenyl ketones with energetically proximate n,π^* and π,π^* triplets, with the latter lower, undergo triplet state hydrogen abstraction with smaller rate constants than those characteristic of n,π^* triplets [56,57]. The decrease in rate constant is proportional to the energy gap between the two triplets [58].

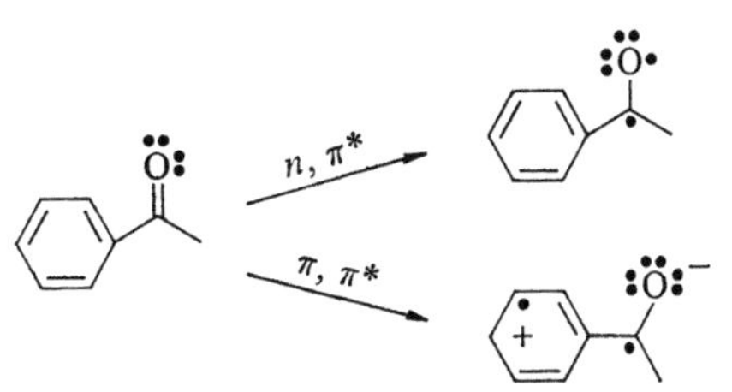

Several explanations have been advanced to explain the radical-like reactivity of excited ketones with π,π^* lowest triplets. Vibronic mixing of the two proximate triplets would make the lowest triplet partially n,π^*-like [57] and might induce some of that state's typical radical reactivity into the resultant mixed state. If the two triplets are close enough, they might equilibrate thermally before decaying, thus allowing a small population of the upper n,π^* triplet to do the hydrogen abstraction, with the unreactive π,π^* triplet serving merely as a reservoir [59].

Inasmuch as naphthyl ketones are not totally unreactive at hydrogen abstractions [54,60], there must be a small amount of reactivity intrinsic to π,π^* states. Such reactivity is actually known for π,π^* olefin triplets [61] and for enone triplets [62]. In the case of phenyl ketones the carbonyl and aryl π-systems are mixed, such that a locally excited carbonyl π,π^* triplet contributes a small fraction to the overall wave function for the π,π^* triplet. Given this intrinsic reactivity, which is some 10^{-4}—10^{-6} that of an n,π^* triplet, the observed reactivity of π,π^* triplets with nearby n,π^* triplets can be expressed as follows, where k_0^{π} is the rate constant

$$k_{obs} = k_0^{\pi} + k_{vib} + \chi_{n,\pi}\, k_H^n$$

$$\chi_{n,\pi} \approx e^{-E_T/RT}$$

describing intrinsic radical-like reactivity, k_{vib} is the rate constant describing any radical-like reactivity resulting from partial n,π^* character, k_H^n is the rate constant for hydrogen abstraction by the upper n,π^* triplet, and $\chi_{n,\pi}$ is the fractional equilibrium population of that state.

The only studies specifically designed to differentiate between vibronic mixing and equilibration as the source of n,π^*-reactivity in ketones with π,π^* lowest triplets have been performed here at Michigan State University. Perhaps the best evidence that equilibration is the major mechanism involves a comparison of intramolecular triplet state hydrogen abstraction by a series of phenyl ketones ($^3n,\pi^*$ lowest) and an analogous series of p-methoxyphenyl ketones ($^3\pi,\pi^*$ lowest).

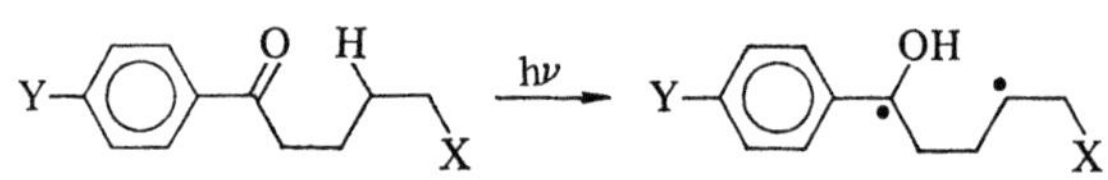

Y=H, CH_3O X=CN, CO_2CH_3, H

The observed rate constants for γ-hydrogen abstraction are decreased in parallel fashion by electron-withdrawing substituents near the γ-carbon for both the benzoyl and anisyl compounds [58,59]. Such a result is expected for equilibration, as long as the substituent does not change the energy gap between n,π^* and π,π^* triplets. However, it is difficult to rationalize how a π,π^* triplet, even with a small fraction of n,π^* character, could become electron deficient at oxygen.

A careful study of polymethyl substituted butyrophenones has revealed that their observed rates of triplet state γ-hydrogen abstraction are no greater than what would be predicted from estimates of ΔE_T, if equilibration of triplets is the *only* source of reactivity [63]. In fact, *meta*- and *para*-methyl ketones are *less* reactive than would have been expected. In these cases, the n,π^* and π,π^* triplets are so close that they presumably undergo maximum vibronic mixing, moving the resultant mixed triplets farther apart and reducing $\chi_{n,\pi}$. The present status of the question is that vibronic coupling is significant only when n,π^* and π,π^* triplets are within 1 kcal of each other and then does not add significant reactivity to the lowest π,π^* triplet.

Since the n,π^* and π,π^* triplets of benzophenone are some 6 kcal apart, the rate constant for hydrogen abstraction from a given substrate by triplet benzophenone is independent of solvent polarity [64]. However, switching from hydrocarbon to alcoholic solvents lowers the triplet state hydrogen abstraction rate constants for phenyl alkyl ketones [58,65], presumably because the lowest triplet becomes largely π,π^* in the polar medium. A similar solvent change lowers triplet hydrogen abstraction rates by p-methyl and p-methoxy phenyl alkyl ketones by a factor of ten [58], suggesting a further separation of 1.4 kcal between the lowest π,π^* triplet and the reactive n,π^* triplet.

B. Inductive Substituent Effects

Triplet state hydrogen abstraction by substituted benzophenones, all of which have n,π^* lowest triplets, reveals that ring substituents exert very little effect on reactivity. A Hammett σ—ϱ plot indicates a ϱ value of only 0.8 — *i.e.*, a two-fold rate reduction for a *p*-methoxy, a three-fold increase for *p*—CN [66]. The direction of this inductive effect is easily understandable in terms of the simplest model for an $n \rightarrow \pi^*$ excitation [2]. Loss of an n electron makes the oxygen somewhat electron deficient.

This electron deficiency is more pronounced when the effects of substituents near the C—H bond are considered. A series of δ-substituted valerophenones display a two order of magnitude variation in rate constants for triplet state γ-hydrogen abstraction, or a ϱ value of —2.0 when σ_I values are used [67]. The ϱ value falls off by a factor of 0.43 for each additional insulating methylene group, suggesting a value of —4.3 for substituents directly attached to the carbon being attacked. Of course, substituents directly on the carbon change the stability of the radical product. Nonetheless, strong electron-withdrawing groups are very deactivating, so that solvents such as acetonitrile, acetone, and acetic acid are very unreactive toward ketone triplets and in fact make excellent solvents for the observation of room temperature phosphorescence of ketones with n,π^* triplets.

C. Bond Strength Effects

As would be expected, rates of hydrogen abstraction are strongly dependent on C—H bond strength. For a given C—H bond, there is very little difference in rate constants among triplet acetone, acetophenone, and benzophenone, all of which possess n,π^* lowest triplets. Activation parameters have been measured for both intramolecular and intermolecular hydrogen atom abstractions. The activation energy for abstraction of an unactivated secondary hydrogen is on the order of 3—3.5 kcal/mole; the activation entropy varies strongly from system to system (see below) [64,68]. This activation energy is comparable to that displayed by *tert*-butoxy radicals, another demonstration of the appropriateness of an alkoxy radical as a model for the n,π^* triplet.

Previtali and Scaiano have attempted to correlate rates of hydrogen abstraction with thermodynamic parameters [69] along the lines of Polanyi [70]. In this approach ($\Delta H^{\neq}$ proportional to ΔH) the exothermicity of hydrogen abstraction varies with the triplet excitation energy and the carbonyl π-bond energy, at constant C—H and O—H bond energies.

$$R_2C{=}O^* + H{-}R' \longrightarrow R_2\dot{C}{-}OH + R'\cdot$$

$$\Delta H = D(R'-H) - E_T + E_\pi - D(O{-}H)$$

The near constancy of k_H values as E_T drops some 10 kcal/mole from acetone to benzophenone is readily explained by the close parallel between E_T and E_π values. Conjugation of the carbonyl with a benzene ring lowers the C—O π-bond energy since the resulting semi-pinacol radical is resonance stabilized. This parallel further supports a simple 1,2-diradical valence bond picture for carbonyl triplets, a picture which implies that the excitation has broken the π-bond and that the n-electron on oxygen is very weakly correlated with the electron in the π-system.

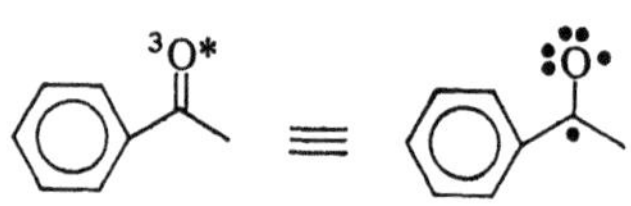

It is at least amusing and perhaps even revealing that the dipole moment of triplet benzophenone is that of a C—O single bond [71]. The measured k_H values for the n,π^* triplets of α-diketones are 2—3 orders of magnitude lower than for monoketones [71a]. It is likely that the low E_T values (~55 kcal/mole) are not compensated for by lowered π-bond energy.

D. Steric Effects

Lewis has reported that rates of bimolecular hydrogen abstraction by triplet phenyl alkyl ketones depend on the size of the alkyl group [53]. Relative rates for attack on 2-propanol by acetophenone, propiophenone, isobutyrophenone, and pivalophenone are 1.0, 0.65, 0.13, 0.035, respectively. It is noteworthy that the *tert*-butyl group does impede hydrogen abstraction significantly. Since triplet ketones act much like alkoxy radicals, it perhaps is time to turn the tables and suggest that the normal assumption that rates of hydrogen abstraction by alkoxy radicals are independent of the alkyl group [27] may need reevaluation.

Intramolecular hydrogen abstraction is dominated by conformational factors which limit the accessibility of most hydrogens in a given molecule. 1,5-Hydrogen atom transfers are by far the most common, as exemplified by the Norrish type II photoelimination reaction. The transition state for γ-hydrogen abstraction can adopt a torsion-free chair conformation [72]. That for β,δ, or more remote hydrogen abstraction would necessarily be destabilized by some of the torsional strain of rings with 5, 7, or more carbons. In fact, the effects of α,β,γ, and δ-methyl groups on rate constants for triplet state γ-hydrogen abstraction indicate that no eclipsing interactions are required in the transition state [72,73].

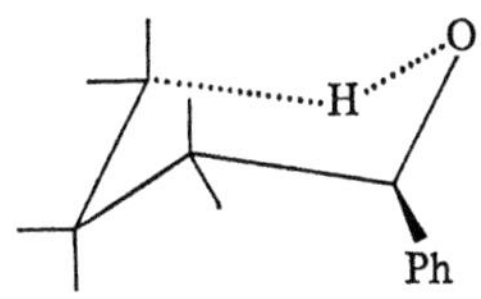

With n-hexyl (and presumably larger) ketones, about 5% δ-hydrogen abstraction competes with γ-hydrogen abstraction [72]. When the δ-carbon is activated by a methoxy group, the ratio increases to 50:50 [74].

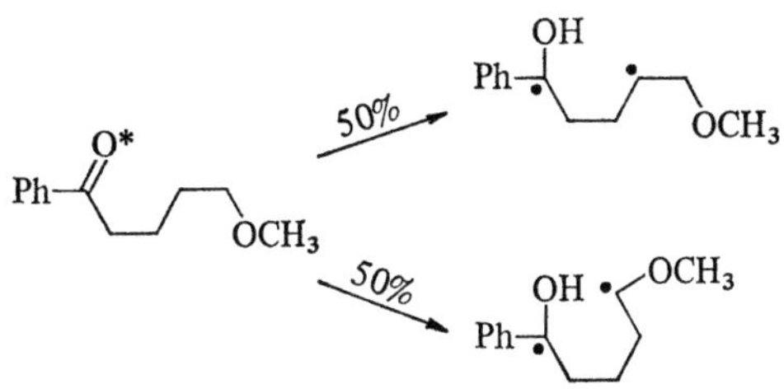

The 20/1 γ/δ ratio, when both carbons are unactivated methylenes, is also displayed by alkoxy radicals [75]. The slower 1,6-hydrogen transfer presumably reflects some of the strain in a 7-membered ring. It must also involve a more negative entropy. Rotations about three C—C bonds are frozen in the transition state, whereas only two C—C rotations need be frozen for γ-hydrogen abstraction.

The entropy argument can be looked at another way, namely as the probability that the triplet can reach a conformation which brings a hydrogen within bonding distance of the carbonyl oxygen. There are of course more conformations involving carbons α through δ than involving just α through γ.

Lewis has reported that γ-hydrogen abstraction in polycyclic ketones is more rapid than in acyclic ketones [68] and concluded that the rate increases are due to the increased number of "frozen" C—C bonds in the reactant. His measured activation parameters certainly support this interpretation.

One factor which has not normally been considered explicitly is the varying distances between γ-hydrogens and the oxygen. At one point it was suggested that hydrogen abstraction had to occur in the nodal plane of the carbonyl π-system since the axis of the n-orbital lies in this plane [76]. It has since been pointed out that there is no very strict such stereoelectronic requirement for hydrogen abstraction [72,73]. In fact, the original evidence [76] cited for a stereoelectronic effect — the lack of triplet state γ-hydrogen abstraction in *trans*-2-propyl-4-*tert*-butylcyclohexanone — ignored the fact that no hydrogens lie anywhere near within bonding distance of the oxygen.

In several steroids, triplet state γ-hydrogen abstraction is efficient even though the γ-hydrogens lie appreciably above (or below) the nodal plane of the

Ketone	$k_{\gamma\text{-H}}$, 10^8 sec^{-1}	*Ea*, kcal	$\Delta S^{\ddagger}$, eu
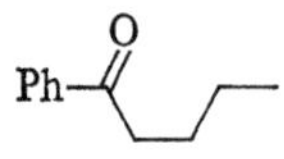	1.2	3.5	−12
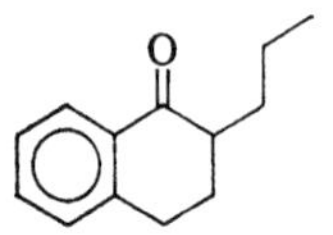	6.0	—	—
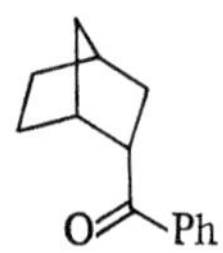	70.	3.7	−4

reactive carbonyl's π-system [77]. The concept of a stereoelectronic requirement may be valid but of limited usefulness. The angular dependence of electron density in a nonbonding p orbital is a $\cos^2\phi$ function. Therefore, for a nearby hydrogen held at $\phi = 60°$, reactivity would presumably be lowered no more than a factor of 1/4. It may indeed be true that a hydrogen held exactly in the plane of the carbonyl π-system ($\phi = 90°$) is unreactive. Sugiyama has so interpreted the lack of light-induced hydrogen abstraction in the following rigid polycyclic ketone.[78] However, other reactions could be happening too fast. What must be determined in order to assess whether a real stereoelectronic requirement exists is the dependence of rates of triplet state hydrogen abstraction on ϕ.

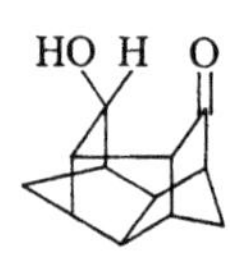

Several different elegant studies have been concerned with the availability of remote hydrogens for intramolecular abstraction by a triplet benzoyl group. Breslow and Winnik first reported remote hydrogen abstraction from *para*-substituted benzophenones [79]. Winnik has shown that the accessibility of

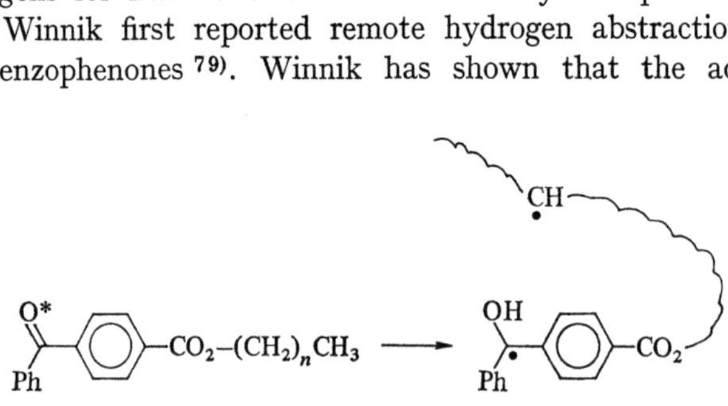

various hydrogens on the remote alkyl chain can be predicted by statistical computations [80]. Measured rate constants for intramolecular hydrogen abstraction also increase as the predicted probability of a favorable conformation increases [81].

Rather regioselective oxidation of steroids is possible by such remote photo-induced hydrogen abstractions [82].

Since rate constants for bimolecular hydrogen abstraction are in the 10^4–$10^6 M^{-1} sec^{-1}$ range, the measured rate of any intramolecular reaction of a flexible molecule includes rotational equilibrium constants. This basic aspect of the kinetics was always assumed in early discussions and was not treated explicitly until recently.

$$\begin{array}{ccccc} \text{unfavorable conformations } (\overset{*}{u}) & \rightleftarrows & \text{favorable conformations } (\overset{*}{f}) & \xrightarrow{k_H} & \text{H-abstraction} \\ \uparrow h\nu & & \uparrow h\nu & & \\ \text{unfavorable conformations } (u) & \rightleftarrows & \text{favorable conformations } (f) & & \end{array}$$

There are three boundary conditions of interest: 1) excited state conformational changes are faster than triplet decay; 2) conformational changes are slower than triplet decay; 3) conformational changes are rate-determining.

When all conformational changes are fast with respect to the rate of hydrogen abstraction, the Winstein-Holness and Curtin principles apply. The observed rate is the sum of rates for each favorable conformation f times the equilibrium percentage of molecules is that conformation. The values of χ_f are the normal Boltz-

$$k_H^{obs} = \sum_f X_f \, k_H^f$$

mann distributions determined by the number and energies of discrete conformers. There is only one triplet lifetime.

$$1/\tau = \sum_u \chi_u \, \tau_u^{-1} + \sum_f \chi_f \, \tau_f^{-1}$$

The lifetimes τ_u of unfavorable conformations are determined by whatever decay reactions — physical and chemical — are available to each conformer. The decay rates τ_f^{-1} of favorable conformations include rates of hydrogen abstraction. Quantum yields for hydrogen abstraction are determined by the relative rates of hydrogen abstraction and of decay, no matter what conformers they arise from. The type II reaction of acyclic ketones falls into this general class. The unit quantum yield for triplet state γ-hydrogen abstraction in many ketones indicates that conformational change is faster than hydrogen abstraction which, in turn, is faster than any other form of decay.

$$\Phi_H = k_H^{obs} \, \tau$$

In ketones with quantum yields lower than unity, some other decay mode is competing with hydrogen abstraction. Cyclobutyl ketones provide nice examples of low quantum yields being due to very low χ_f values. The rate of decay of triplet benzoylcyclobutane is $5 \times 10^5 sec^{-1}$ in benzene [83]; the quantum yield for type II elimination and cyclization is only 0.03 and varies sharply with substitution on the benzene ring [84]. Hydrogen abstraction can occur only when the benzoyl group is in the pseudo-axial conformation *a*.

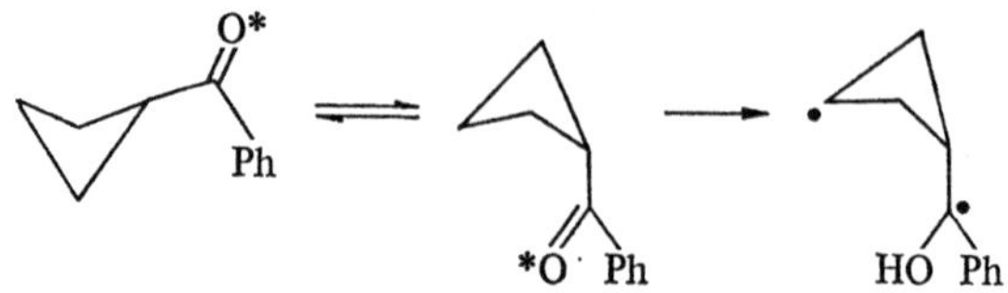

e *a*

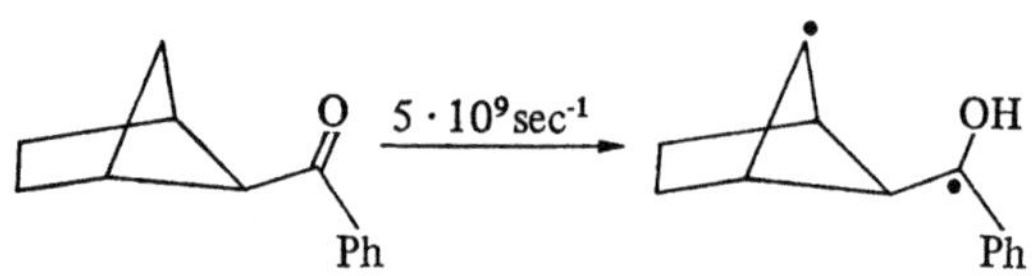

Since exo-5-benzoylbicyclo[2.1.1]hexane, a good model for *a*, undergoes triplet state γ-hydrogen abstraction very rapidly [85], the low apparent reactivity of triplet benzoylcyclobutane indicates that only $< 0.001\%$ of the triplets are in the pseudoaxial conformation at equilibrium.

Lewis has provided the nicest example of a system in which excited state reactions are faster than conformational change, such that excited state reactions are determined by ground state conformational preferences. Excitation of 1-benzoyl-1-methylcyclohexane yields two discrete triplets: one with a lifetime of 10^{-7} sec which undergoes only α-cleavage to radicals; and one with a much shorter lifetime which undergoes only γ-hydrogen abstraction and cyclization [86]. The 10^5 sec^{-1} rate of cyclohexane ring inversion is too slow to equilibrate excited conformers with the benzoyl group equatorial and axial.

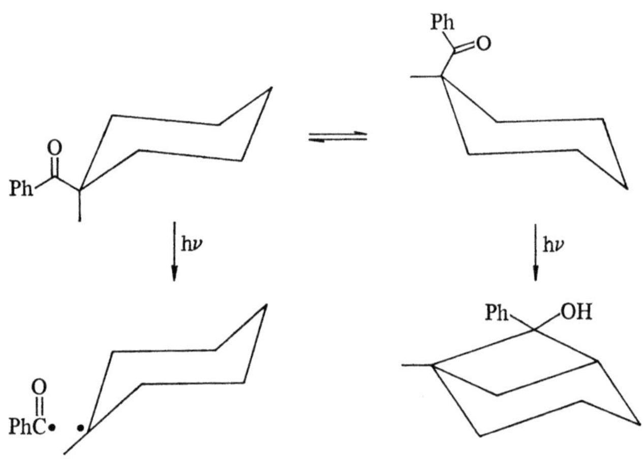

We have recently discovered at Michigan State University that the well-known photoenolization of *ortho*-alkyl ketones [87] involves two distinct rotamers [88]. The *syn*-triplet enolizes very rapidly ($> 10^9$ sec^{-1}), the *anti*-triplet so slowly ($\sim 10^7$ sec^{-1}) that γ-hydrogen abstraction can compete. The rate of enolization of the *anti*-triplet is apparently determined by the necessary rotation into the *syn* conformation, since it is independent of substituents on the benzene ring and of the nature of the *ortho*-alkyl group.

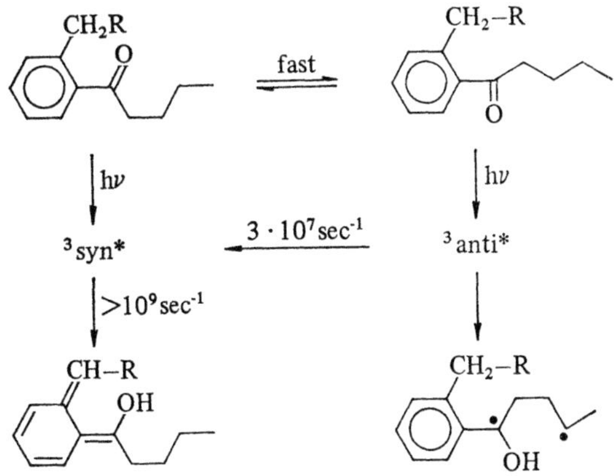

The type II reaction of these *ortho*-alkyl ketones comes only from the ~20% of the ground states that are in the *anti* conformation. Enolization, on the other hand, comes from both conformations. That from the *syn* conformation apparently is so fast that it is kinetically "static"; that from the *anti* conformation is dynamic.

E. Triplet vs. Singlet Reactivity

The relative reactivities of n,π^* singlets in hydrogen abstraction reactions has not yet been defined as precisely as for n,π^* triplets. There is little doubt that n,π^* singlets do abstract hydrogen atoms, since the unquenchable type II cyclization of compounds such as cyclohexenyl acetone involve allylic rearrangement [89,90].

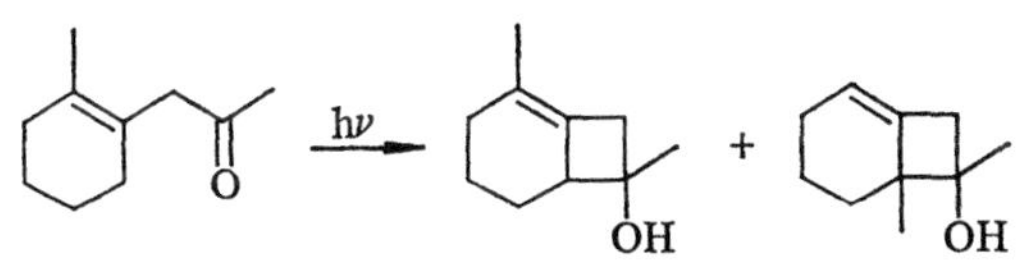

The singlet state type II reaction of aliphatic ketones proceeds very inefficiently [90] and in a highly stereospecific fashion [91], as opposed to the highly efficient [9], nonstereospecific triplet reaction [91]. Although an extremely short-lived singlet diradical has been invoked [90,91], it has been clear for some time that the situation may be more complicated [9].

Heller first proposed that electronic energy might be transferred into vibrational stretching energy of a C—H bond [92], with further partitioning of the hot product into a pair of radicals (or diradical) or to a relaxed ground state. This proposal was perhaps the first specific suggestion in line with Hammond's latest postulate, that chemical reactions of excited states are special forms of radiationless decay [93]. It was suggested in 1971 that singlet (but not triplet) state type II reactions may be one example of such coupled chemistry and radiationless decay [9]. Since then, understanding of radiationless decay has improved considerably and it appears that a reaction coordinate for chemical change can indeed promote radiationless decay of excited singlets. The following picture of excited ketone

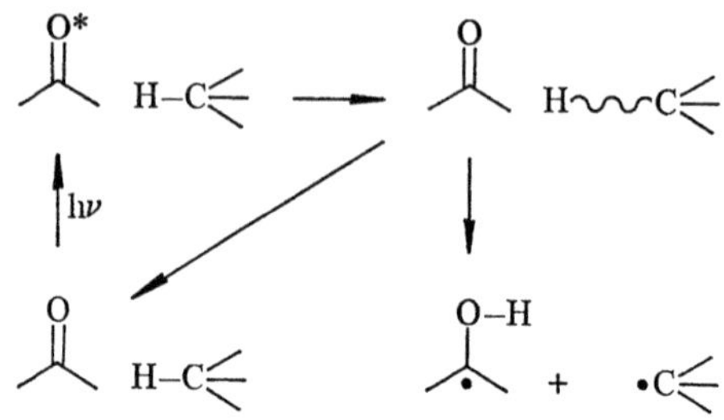

hydrogen abstraction can be pieced together from the recent papers of Michl [94] and of Salem [29,95], with only a few embellishments by the present author. The triplet ketone can proceed smoothly to a triplet diradical state, the efficiency being limited only by the rates of any competing reactions [67,96]. As the CIDNP

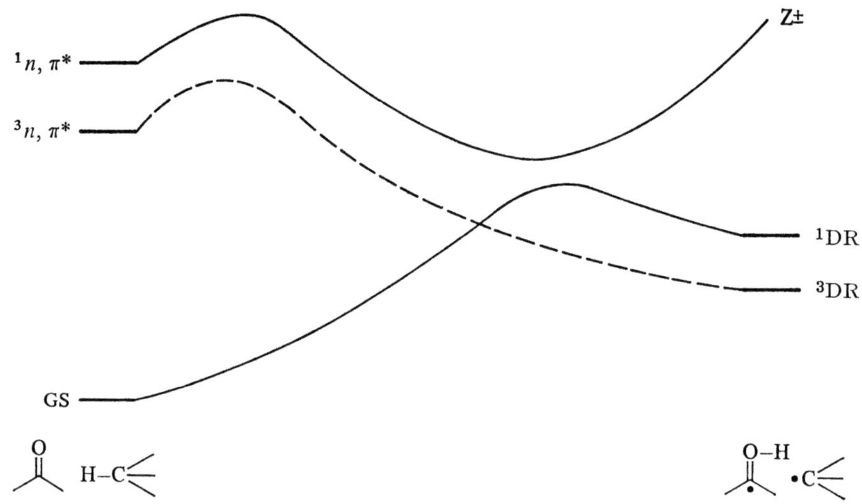

phenomenon shows, spin inversions in diradicals and radical pairs is rapid enough that triplet diradicals can go on easily to singlet products. In fact, the potential energy diagram indicated by the dashed line can be considered to arise from the normal combination of two potential wells. The only difference between triplet

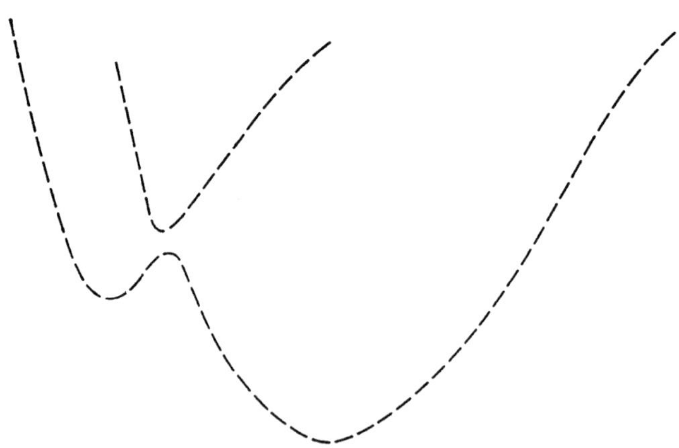

state hydrogen abstraction and hydrogen abstraction by a ground state radical is that, in the former, the potential wells for reactant and product lie on a common *triplet,* rather than ground state, hypersurface.

Simple symmetry considerations indicate that the n,π^* singlet of a carbonyl correlates with the singlet diradical product, whereas the ground state correlates with a zwitterionic divalent species [29]. Since a plane of symmetry is not maintained in most hydrogen abstractions, there is an avoided crossing between the excited surface and the ground state surface [29]. Hydrogen abstraction therefore requires a radiationless decay which can occur easiest at the point of smallest energy difference between the two surfaces [94]. This minimum in the excited surface occurs at a reaction coordinate where the C-H bond is stretched but not yet fully broken [94]. The sudden change of electronic energy into vibrational energy populates the ground state surface on a maximum, from which redistribution of vibrational energy can lead on to diradical product or back to ground state reactant. In other words, singlet hydrogen abstraction is inherently inefficient [94,97], even if no other reaction of the excited singlet can compete, exactly the situation in singlet state type II eliminations [9,90]. The potential energy diagram for excited singlet hydrogen abstraction is different from that for normal radical hydrogen abstraction only in the involvement of a lower energy hypersurface in the latter stages of the reaction. Whatever activation barrier must be surmounted results from the normal avoided crossing of two potential wells of the same symmetry on the same singlet hypersurface. (These last two sentences are the present author's contribution to the problem.)

These recent theoretical studies seem to support the early suggestion [9] that the inefficiency of singlet state type II elimination may be explained by Heller's electronic-to-vibrational energy transfer. What was missing earlier was a good explanation of *why* this form of radiationless decay should occur. It is still disturbing, however, that the experimental evidence regarding bimolecular hydrogen abstraction by n,π^* singlets is somewhat confusing.

The weak quenching of alkanone fluorescence by several alcohols [98] correlates better with their ionization potentials than with their C—H bond strengths. Several β,γ-unsaturated ketones and aldehydes are photoreduced by stannanes even in the presence of triplet quenchers [32]. Aliphatic ketones are photoreduced in 100%-quantum yield by tributylstannane but the reaction is quenched totally by conjugated dienes [99,100]. The stannane does not quench the fluorescence intensity of acetone [101] but does shorten the singlet lifetime of adamantanone [98]. The bimolecular rate constant which describes the ability of stannane to shorten the singlet lifetime of adamantanone is 5×10^8 M^{-1} sec^{-1}, similar to the measured rate constant for triplet state hydrogen abstraction from stannane [99]. If one assumes some error in the fluorescence measurements, this singlet quenching must either produce radicals in 100% efficiency or be a heavy atom induced intersystem crossing. otherwise the quantum yields for photoreduction in the absence of triplet quenchers could not be unity. Dienes might well intercept hydroxy radicals [98] and thus *chemically* quench singlet state photoreduction, but it is unlikely that naphthalene would do so. We have found a quantum yield of 0.07 for isopropyl alcohol formation in 1 M acetone-0.4 M tributylstannane-3 M 1-methylnaphthalene mixtures [102], conditions [8] in which all of the light is absorbed by the naphthalene, 98% of the naphthalene singlets transfer their energy to acetone (stannane does not quench naphthalene fluorescence), and 99% of the acetone triplets are quenched by naphthalene. With the rate constant for intersystem crossing in acetone equal to 5×10^8 sec^{-1}, the effective rate constant for singlet hydrogen abstraction is only 1×10^8 M^{-1} sec^{-1}. The evidence about the ability of n,π^* carbonyl singlets to abstract a hydrogen atom from very reactive hydrogen donors such as stannanes is contradictory and no firm conclusion can yet be reached about this important comparison between singlet and triplet reactivities.

F. Radicals and Diradicals

Excited states are essentially diradicals [2,97]. Their early-noted [2] propensity to yield diradical products was always intuitively reasonable and is now understood more fully in terms of state correlation diagrams [97]. For triplets, diradical formation can occur without spin inverson and is particularly favorable since it allows the parallel spins to get farther apart [96]. In early discussions the rate of triplet→singlet spin inversion in primary diradical photoproducts was always tacitly assumed to be fast enough that ground state (singlet) product formation could proceed without further ado. However, the stereospecificity of product formation from cyclic azo compounds depends so strongly on whether direct (singlet) irradiation or triplet sensitization is employed [103] that it is widely believed that the rate of triplet→singlet spin inversion is rate-determining in the reactions of triplet-derived 1,3- and 1,4-diradicals. Since the type II reaction allows the production of 1,4-diradicals of great structural variety, it has been possible to learn much about diradicals from careful study of this reaction.

The ability of thiols to trap the diradical intermediates [104] and the partial rearrangement of α-allylbutyrophenone to 2-phenyl-2-norbornanol [105] strongly

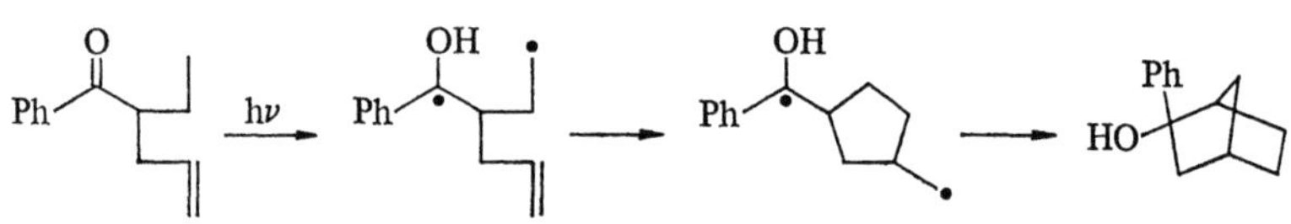

suggest that the rate constants for cyclization and for cleavage of the diradicals are on the order of 10^6 sec^{-1}. It would be surprising if spin inversion were this slow, unless through bond coupling [106,107] produced an unusually large exchange interaction. There is considerable evidence that substituents which increase rotational barriers alter product ratios from triplet-derived diradicals [28,72,73,108]. Cyclobutanols formed from optically active ketones with asymmetric γ-carbons always retain some optical activity [96,109,110]. Therefore, triplet-derived diradicals do not live long enough for complete rotational equilibration to occur. It has been proposed that normal activation barriers, not rates of spin inversion, determine the lifetimes of diradicals [107] with strong through-bond coupling inhibiting rotation in singlet, but not triplet, 1,4-diradicals [111].

The actual reactions of the diradicals and radical pairs formed by triplet state hydrogen abstraction are those expected of radical pairs: coupling (cyclization);

disproportionation; cleavage (1,4-diradicals); radical rearrangements; further hydrogen abstraction. Not enough structural variations have been studied in order to allow reliable predictability of the actual partitioning of a given diradical or radical pair. Steric hindrance clearly decreases coupling. Thus pivalophenone is photoreduced to the alcohol rather than to pinacol [53]. However, *tert*-alkyl ketones with γ C—H bonds undergo type II cyclization in preference to elimination.[47] It has been suggested [74,108] that α-substituents impede rotation of the diradical into a conformation required for cleavage, one with the 2,3 C—C σ bond parallel to both *p* orbitals at carbons 1 and 4[112]. Such a stereoelectronic requirement for cleavage seems indicated by both theoretical [106] and experimental work [9,113]. Since type II cyclization is one of the best synthetic entries to four-membered rings, the structural parameters which favor cyclization *vs.* cleavage are of some interest.

Inductive effects of substituents on diradical behavior are most pronounced in the disproportionation reaction. With phenyl ketones, electron-donating substituents in the *para* position and electron-withdrawing substituents in the γ-position enhance the probability for product formation from the diradical intermediate, at the expense of disproportionation. This disproportionation is now seen as a normal, low activation energy, ground state reaction [29,94]. It has been

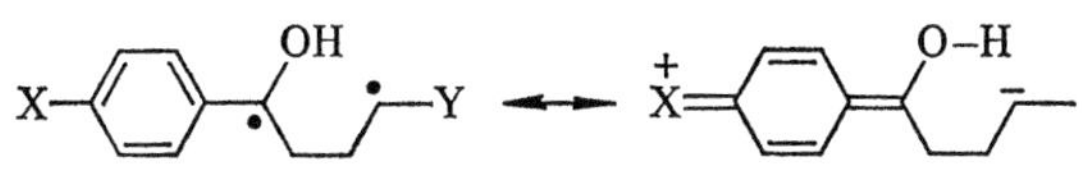

suggested that partial zwitterionic character in the (singlet) diradical might affect disproportionation, since the singlet ground state and a zwitterion have the same symmetry [97]. Experimentally, substituents which make the hydroxyl proton more acidic and the γ-radical site more easily oxidizable favor disproportionation [108]. This phenomenon is an example of the now well documented electron donating character of alkyl radicals in hydrogen atom abstraction reactions [114]. It is not a zwitterionic contribution to the diradical but the *correct*

zwitterionic contribution to the transition state for disproportionation that lowers type II quantum yields.

The mere fact that so much diradical disproportionation occurs in triplet type II reactions reflects the acidity of the hydroxyl proton. The ability of even very weak Lewis bases to suppress this disproportionation [115] is further evidence. It often happens that added Lewis base causes P_p to equal unity, such that quantum yields are determined solely by competing triplet reactions. Unfortunately, in sterically crowded ketones, such is not always the case.

Since α-hydroxyradicals are the primary photoproducts in any hydrogen abstraction by ketones, their strong reducing power should be noted. The photoreduction of benzophenone in secondary alcohols provided the first example of the reducing capabilities of hydroxyradicals [116].

$$Ph_2C = O^* + R_2CH\text{—}OH \longrightarrow Ph_2\dot{C}\text{—}OH + R_2\dot{C}\text{—}OH$$

$$R_2\dot{C}\text{—}OH + Ph_2C = O \longrightarrow R_2C = 0 + Ph_2\dot{C}\text{—}OH$$

The second reaction above was verified by thermal production of α-hydroxyradicals [117]. There are now many reports of *chemical* sensitization by ketones, wherein a α-hydroxyradical photoproduct reduces a ground state of another compound. α-Diketones [118], α,β-unsaturated ketones [119], and imines [120] all undergo such chemically sensitized photoreductions.

Finally, there is one aspect of the photoreduction of phenyl ketones which has been resolved only recently. It is now well established that benzyl radicals couple to form 1-alkylidene-2,5-cyclohexadienes as well as bibenzyls, the ratio depending primarily on steric factors [121]. Schenck first showed that the semipinacol radicals involved in the photoreduction of benzophenone also form such products [122]. These compounds absorb very strongly in the near UV and often

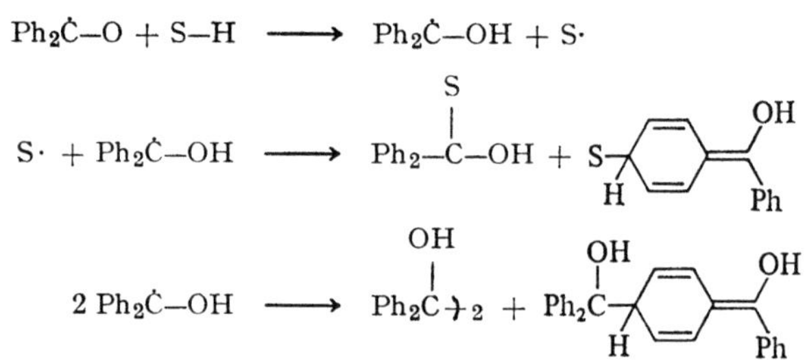

function as internal filters lowering quantum efficiency. Their presence also interferes with UV analysis of ketone disappearance, although they do slowly disappear by as yet undefined reactions. In particular, they react rapidly with oxygen.

I concluded in 1969 that only a few per cent of ring coupling occurs in benzophenone photoreduction but that the triene products effectively quench all high energy triplets present, both those of benzophenone and of several aromatic quenchers [123]. Steel has now verified this conclusion by flash spectroscopic studies [124]. Despite the fact that the UV spectra of irradiated solutions in which phenyl ketones have undergone photoreduction are dominated by triene absorption [125], FT NMR of an irradiated benzophenone solution indicates that benzpinacol is the predominant (only readily detectable) product [124]. The earlier suggestion [125] that trienes were the first-formed products and that they slowly rearrange to pinacols is not true.

V. Charge Transfer

A. General Considerations

All Excited states are easier to oxidize and reduce than are ground states. In particular, carbonyl compounds undergo one-electron reductions fairly readily to yield relatively stable ketyl radical-anions. The triplet excited ketone is

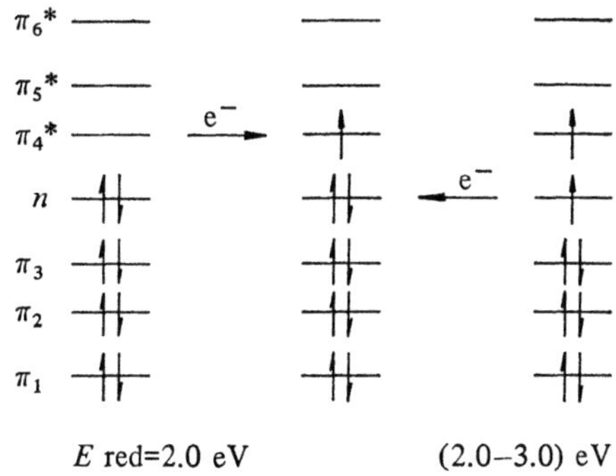

easier to reduce than the ground state by the triplet excitation energy, as the figure illustrates. Simple phenyl ketones are reduced polarographically at approximately —2 ev relative to a standard calomel electrode. The excited ketone is thus reduced at a 1V positive potential relative to sce.

Since excited ketones are relatively strong oxidizing agents, it might be expected that they would participate in a wide variety of one-electron redox reactions. In fact, they do not seem to. Instead, they form charge transfer complexes with almost all electron donors. These complexes are probably exciplexes, although there is no evidence that they either dissociate back to excited ketone or phosphoresce. These complexes do undergo facile chemical rearrangement and radiationless decay. Two important reaction classes — the cycloaddition of triplet ketones to olefins and the reduction of triplet ketones by amines — apparently involve such intermediates. The rest of this section consists of discussions of these two reactions followed by a discussion of CT quenching in general.

B. Cycloadditions to Olefins

1. Mechanism

It has been thought for some time that the interactions of triplet ketones with olefins involve a competition between hydrogen abstraction from, energy transfer to, and cycloaddition to the double bond [126]. Cycloaddition has generally been considered to proceed via a diradical intermediate. At first, the only evidence for a diradical was that the regioselectivity of oxetane formation often — but not always — is such as would be expected from the relative stabilities of the possible diradical intermediates [126]. Moreover, n,π^* ketone triplets are known to act like alkoxy radicals, which add to olefins, albeit less efficiently than do most other radicals [27].

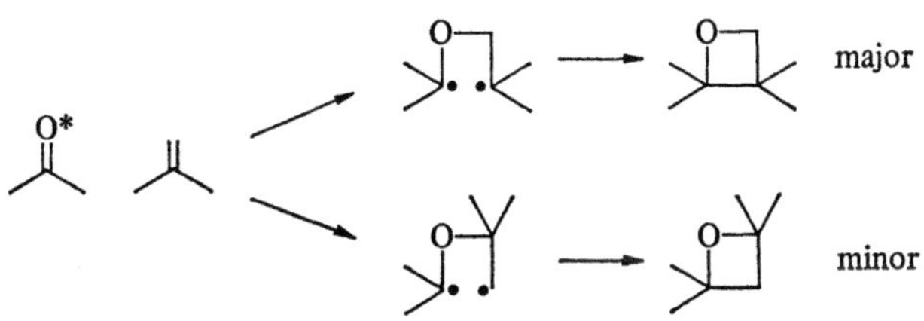

Carless has reported a careful study of the photoreaction between acetone and 2,3-dimethyl-2-butene [127]. Together with photoreduction products and oxetane are formed two alcohols which can be explained simply as internal disproportionation products of the diradical.

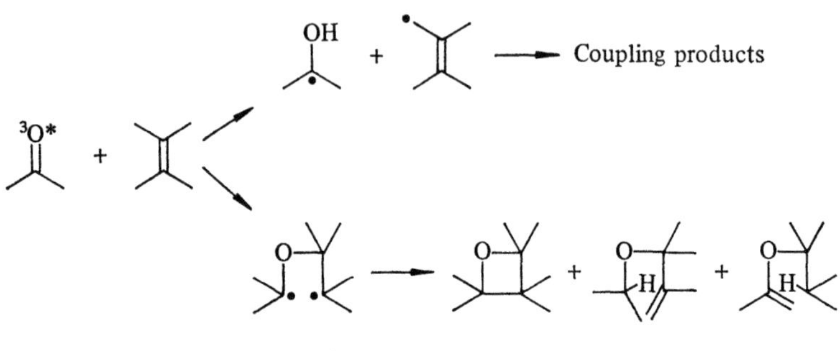

Earlier, Saltiel provided indirect evidence for a diradical intermediate by noting variations in the photosensitized *cis/trans* ratio of 2-alkenes [128]. Originally, sensitized isomerization of olefins was thought to involve only energy transfer

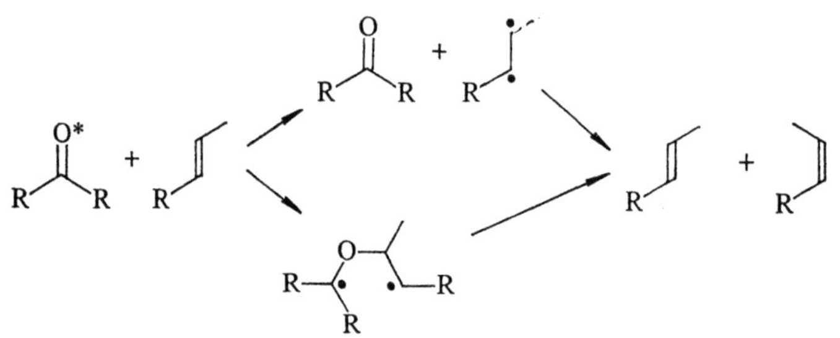

from the triplet ketone. However, any real 1,4-diradical intermediate would certainly cleave at least partially to two olefins (the same diradical reaction responsible for type II photoelimination). The *cis/trans* alkene ratio so formed in general could be different from that formed by decay of the alkene triplet.

Photoöxetane formation is quite inefficient, a fact which usually points to the presence of an intermediate which can partially revert to ground state reactants. Cleavage of the diradical must be responsible for some of the inefficiency in oxetane formation [129]. However, in the past few years convincing evidence has appeared that a CT complex precedes the diradical [130,131]. The two most telling pieces of evidence are the relative reactivities of different alkenes [130] and the absence of any measurable secondary deuterium isotope effect on quenching rate constants [131]. Relative quenching rates of sterically uncrowded olefins are proportional both to the ionization potentials of the donor olefins [130] and to the reduction potentials of the acceptor ketones [131], as would be expected for a CT process. Inasmuch as n,π^* triplets resemble electron-deficient alkoxy radicals, such substituent effects would also be expected on direct radical addition of triplet ketone to olefin. However, radical addition would yield an inverse isotope effect (in, say, 2-butene-2,3-d_2) and would be faster to 1,1-dialkylethylenes than to 1,2-dialkylethylenes, in contrast to the actual observations.

The regioselectivity in oxetane formation is probably determined primarily by the orientation of the CT complexes. Additions of α-fluoroketones, which have such low reduction potentials as to be particularly hot [132] and therefore potentially unselective acceptors, do proceed with negligible regioselectivity [126]. The actual product ratio is quite complex kinetically because of all the various competing reactions of four separate intermediates.

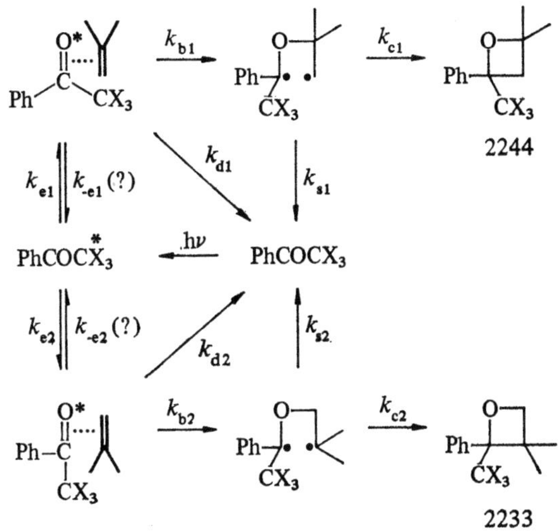

$$\frac{2233}{2244} = \frac{k_{e1}}{k_{e2}} \cdot \frac{k_{b1}\,(k_{-e2} + k_{d2} + k_{b2})}{k_{b2}\,(k_{-e1} + k_{d1} + k_{b1})} \cdot \frac{k_{c1}\,(k_{c2} + k_{s2})}{k_{c2}\,(k_{c1} + k_{s1})}$$

For example, the change in the 2233/2244 ratio from 9:1 for X=H to 1.5 for X=F [126] involves possible changes in 12 independent rate constants, even though the simple inductive argument noted above would hope that the biggest change is in k_{e1}/k_{e2}. Steric features of both reactants may produce large variations in all rate constants. For example, as steric crowding increases, k_b and k_c may well decrease while k_e, k_d ,and k_s may increase. In fact, some olefins quench triplet ketones so efficiently without yielding products in significant yields [130] that one may infer that k_d often is larger than k_b.

The orientation of the two molecules in the CT complex is not well understood. Turro has presented evidence that substituents on the ketone in the nodal plane of the carbonyl π-system sterically hinder CT quenching of the triplet ketone [133]. The effect is not large; but given that the approach between donor and acceptor in CT complexes is not as close as required for covalent bonding, a large effect is not needed for one to infer that the donor orbital probably overlaps with the half-vacant *n*-orbital. Inasmuch as the olefin has radical-catonic

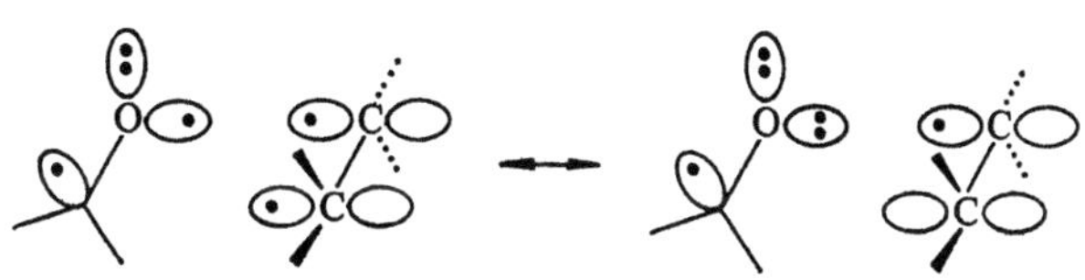

character in the CT complex, most of the charge should be situated on the more-substituted carbon of the double bond. Complexation to an unsymmetrical olefin such as isobutylene might involve two orientations with charge-separated resonance contributors as shown below. It is possible that in a fluoro-substituted

ketone, the CT complex involves more electron transfer (*i.e.*, the charge separated form contributes more to the total) than in a less electron-demanding ketone. The resulting extra charge on the dimethyl-substituted carbon would then force the complex into a very asymmetric orientation with positive and negative charges as close together as possible, such that collapse to the less stable diradical could occur readily.

As mentioned above, if the CT complex is an exciplex it might revert to monomeric excited ketone. Caldwell's isotope studies rule out this possibility, since if complex formation were significantly reversible, the observed rate constant for quenching would be proportional to k_b and thus should show an isotope effect.

2. Singlet vs. Triplet Reactivity

Oxetane formation is not only a triplet reaction. Singlet n,π^* states of aliphatic ketones add stereospecifically to electron-deficient olefins, whereas their triplet states transfer energy to these olefins [134]. With olefins such as 1-methoxy-1-butene, oxetane formation takes place from both singlet and triplet states [135].

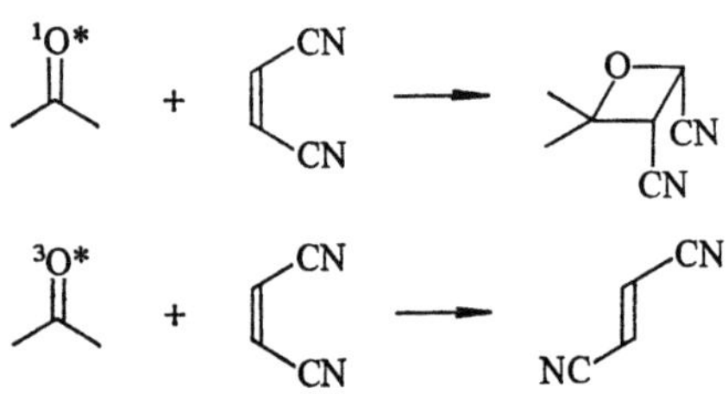

Yang reported that quenching of excited singlet alkanones by alkenes and by dienes is so dependent on the quencher's ionization potential that a CT process is likely [136]. The rate constants which describe quenching of singlet and triplet ketones by a given olefin are very similar, as would be expected for similar processes.

3. n,π^* vs. π,π^* Reactivity

There has been surprisingly little work done to differentiate the relative reactivities of n,π^* and π,π^* triplets in oxetane formation. Triplet fluorenone does add to ketenimines [137]. Naphthyl [138] and biphenylyl [126] ketones also form oxetanes. It is not known whether the lowest π,π^* triplets are responsible or whether an upper n,π^* state reacts, as observed for 9-anthraldehyde [139]. It is entirely possible that olefins are active in forming CT complexes with π,π^* ketones but that the complexes rearrange very inefficiently to diradicals and oxetanes.

4. α,β-Unsaturated Ketones

The cycloaddition of enones to olefins is a reaction of considerable synthetic interest [140]. Oxetane formation and cyclobutane formation are sometimes competitive [141], but the latter reaction is the more common. The photodimerization of enones [142] is a special case of such cycloaddition. It has been shown that triplets are involved in these cycloadditions, since intersystem crossing quantum yields are unity [143] and cycloaddition is totally quenchable by triplet quenchers. Careful kinetic analysis indicates an intermediate which can partially revert to ground state reactants, since quantum yields are lower than unity even when extrapolated to infinite substrate olefin concentration. That a diradical is

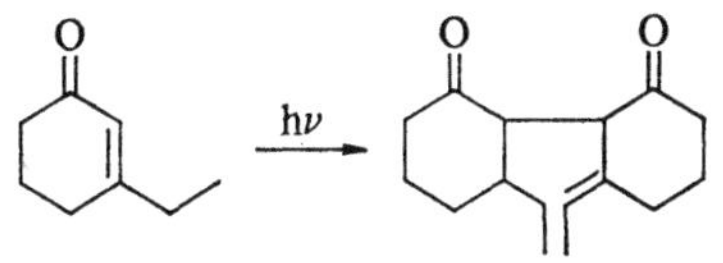

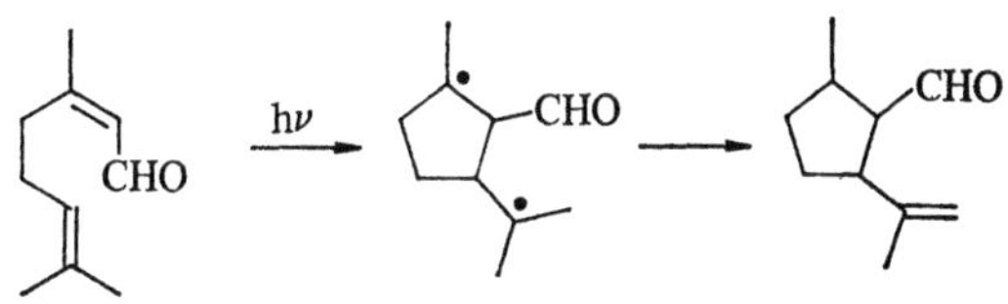

involved seems very likely, as judged by the disproportionation products observed in several reactions [144–146]. Moreover, there is no stereospecificity in the addition [140,147].

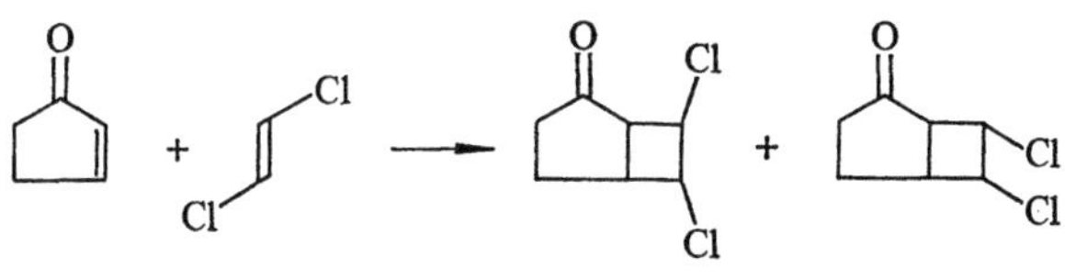

It is assumed that a CT complex precedes diradical formation [140,143,146], primarily because of the regioselectivity of addition to unsymmetrical olefins. If different complexes are indeed involved, the final product ratio is a complex function of many competing rates. For example, let us assume the *least* complicated scheme as shown below for cyclopentenone.

The final equation is included not merely as a display of kinetics prowess. Several authors have pointed out that product ratios are not a simple function of k_H/k_T; the equation ought to make that fact indelibly clear. Moreover, the scheme shown ignores many likely complications. Reversibility of complex formation is ignored. The stereochemistry of H and T are ignored. The diradicals

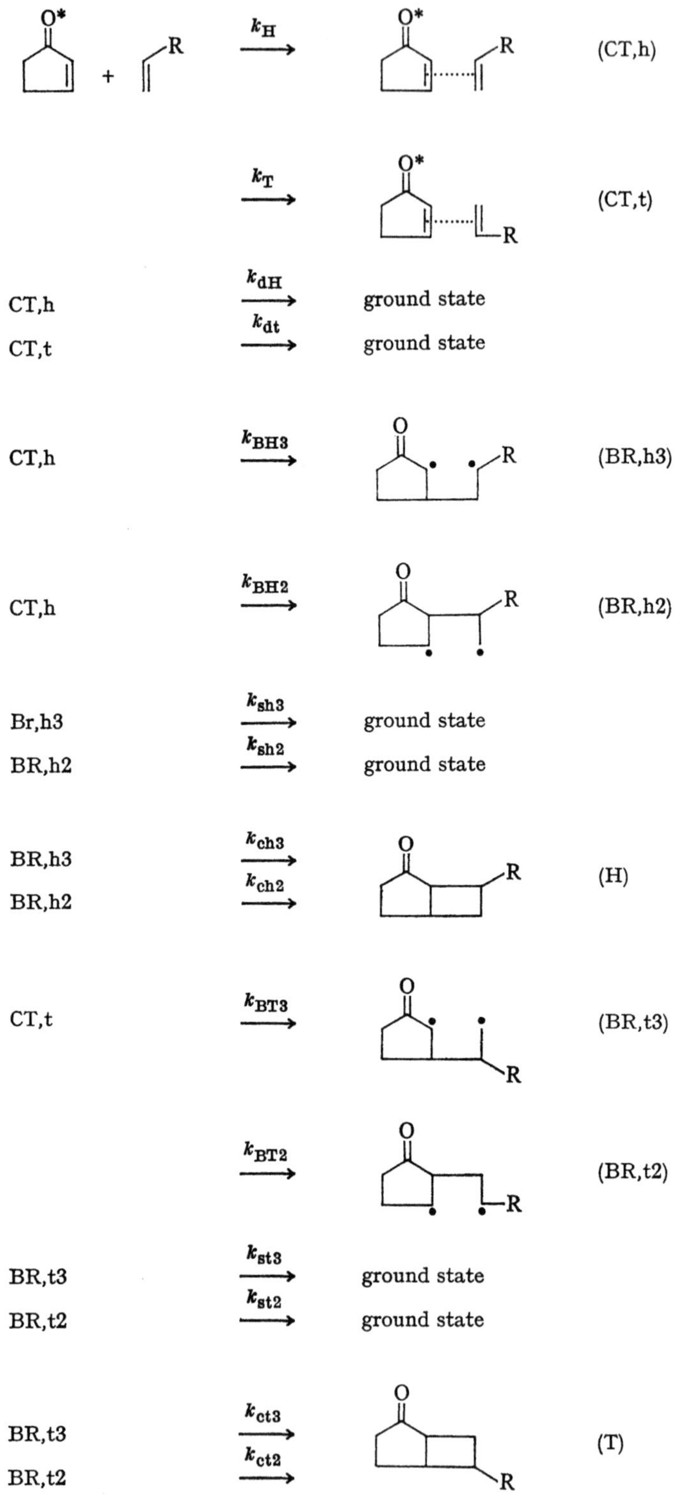

O*
R
+
k_H
(CT,h)
k_T
(CT,t)
CT,h
k_{dH}
ground state
CT,t
k_{dt}
ground state
CT,h
k_{BH3}
(BR,h3)
CT,h
k_{BH2}
(BR,h2)
Br,h3
k_{sh3}
ground state
BR,h2
k_{sh2}
ground state
BR,h3
k_{ch3}
BR,h2
k_{ch2}
(H)
CT,t
k_{BT3}
(BR,t3)
k_{BT2}
(BR,t2)
BR,t3
k_{st3}
ground state
BR,t2
k_{st2}
ground state
BR,t3
k_{ct3}
BR,t2
k_{ct2}
(T)

$$\frac{H}{T} = \frac{k_H \left[\dfrac{k_{BH3} \cdot k_{CH3}}{(k_{BH3} + k_{BH2} + k_{dH})(k_{CH3} + k_{SH3})} + \dfrac{k_{BH2} \cdot k_{CH2}}{(k_{BH3} + k_{BH2} + k_{dH})(k_{CH2} + k_{SH2})} \right]}{k_T \left[\dfrac{k_{BT3} \cdot k_{CT3}}{(k_{BT3} + k_{BT2} + k_{dT})(k_{CT3} + k_{ST3})} + \dfrac{k_{BT2} \cdot k_{CT2}}{(k_{BT3} + k_{BT2} + k_{dH})(k_{CT2} + k_{ST2})} \right]}$$

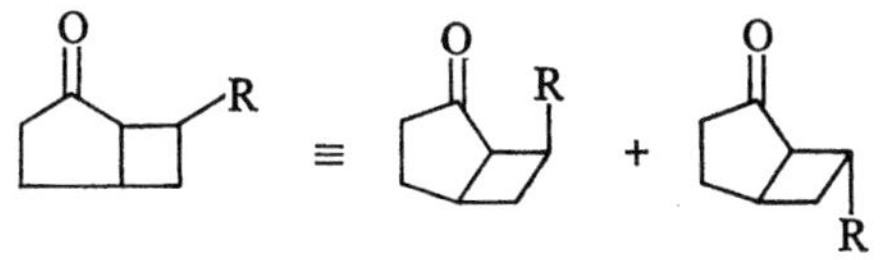

labelled h3 and h2 may well couple to different ratios of the two H isomers. The worst possibility is that each CT complex may well collapse to both diradicals, albeit in potentially different ratios.

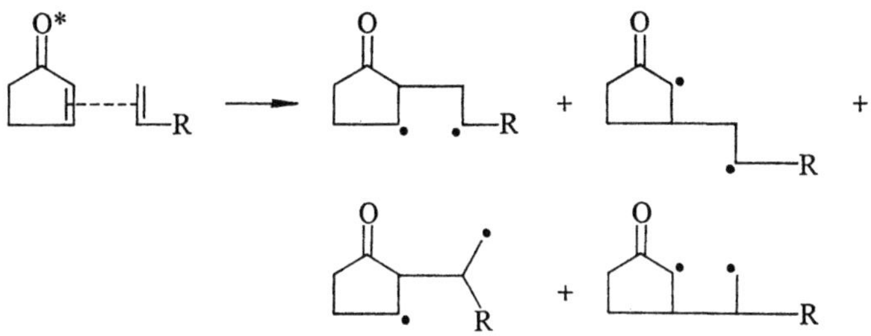

There is no clear evidence which carbon of the double bond forms the first covalent bond in diradical formation. In fact, there is evidence from rearrangement products that each does [148].

Yet another puzzle exists with respect to the nature of the reacting triplet. It is now widely accepted that the *n,π** and *π,π** triplets of enones lie very close together [140,143]. One can argue, mostly by analogy, that the *π,π** triplets are responsible for cyclobutane formation and the *n,π** triplets for oxetane formation and photoreduction. The large rate constants for reaction of triplet cyclopentenone with olefins [140] indicate that the lowest triplet, which is probably mostly *π,π**, is the reactive species in cycloaddition. Solvent changes which would be expected to destabilize the *π,π** triplet do elicit more reduction and oxetane formation [141,149].

It is fortunate that the kinetic complexities and mechanistic ambiguities in enone cycloadditions, which promise to remain with us for some time, have not prevented some really elegant synthetic applications [140,142,150,151].

C. Photoreduction by Amines

1. Evidence for CT Mechanism

Cohen has recently reviewed the large amount of research performed on photoreduction by amines [152]. The overall reaction is different for tertiary amines than for primary and secondary, since radicals with the structure $R_2\dot{C}$–NH–R are able to reduce ground state ketone in the same way that α-hydroxy radicals do.

$$R_2C{=}O + R_2\dot{C}{-}NH{-}R \longrightarrow R_2\dot{C}{-}OH + R_2C{=}NR$$

Photoreduction by amines differs from photoreduction by alcohols in two respects: quantum yields are always lower than maximal and rate constants for amine quenching of triplet ketones are very large. These two facts led Cohen [153] and Davidson [154] to suggest that amines react with excited carbonyl compounds by electron transfer followed by proton transfer from the amine radical cation.

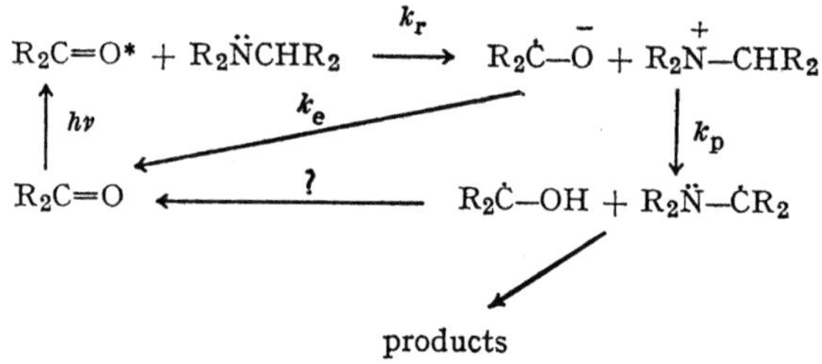

It was later pointed out that a charge transfer rather than electron transfer process was more likely [155]. Now most people would call the product formed by interaction of a triplet ketone and an amine an exciplex, although there is no evidence for any low energy phosphorescence or for reversal to monomeric excited ketone.

The best evidence for a CT process rather than direct hydrogen abstraction involves the values of k_r: *s*-butyl- and *tert*-butylamine display much the same value [156]; triethylamine and *tert*-butyldimethylamine are equally reactive and some 50 times more so than primary amines [155]. Thus the rate constant for reaction is independent not only of the type of C—H bond α to the nitrogen but also of the presence or absence of α-hydrogens. Such evidence demands that abstraction of an α-hydrogen not be involved in the rate-determining quenching reaction. Moreover, the relative reactivity of amines (tertiary > secondary > primary) is proportional to the ease with which they are oxidized.

Although all amines quench triplet ketones rapidly, the efficiency with which the interaction produces radicals varies considerably.

The dependence of k_P on the type of α-hydrogen is not known, nor can variations in k_e be predicted easily. Mataga has studied amine quenching of triplet benzophenone by flash spectroscopy. N,N-dialkylanilines are the only amines which actually yield radical ions, and then only in polar solvents [157]. He suggests two competing decay modes of the exciplex. Monoalkylanilines and tertiary aliphatic amines in any solvent and dialkylanilines in nonpolar solvents yield only radicals, presumably from the exciplex. Even though the oxidation potentials of tertiary aliphatic amines are so low that they quench triplet ketones at rates

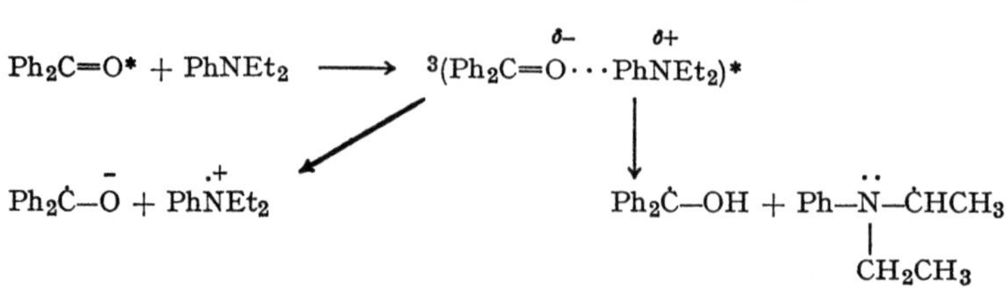

close to diffusion-controlled, full electron transfer does not occur easily until the amine has a very low oxidation potential.

The high rate constants for chemical quenching of triplet ketones by amines provide two sidelights of considerable importance. Photoinitiation of polymerization has received widespread and varied industrial applications. One problem is that many vinyl monomers quench triplet ketones very rapidly either by charge transfer or energy transfer mechanisms, without forming any radicals. Most solvents cannot compete with the olefins for the triplet ketone. However, triethylamine quenches at rates close to diffusion-controlled, so that radical formation and polymerization initiation are quite efficient [158)].

Second, it is conceivable that amines quench triplet ketones before spin-lattice relaxation takes place within the three triplet sublevels [159)], which are populated unevenly [160)]. In that event, radicals can be produced with their electron spins polarized. The CIDEP phenomenon [161)], whereby EPR emission is observed upon irradiation of ketones and very reactive substrates, may involve this mechanism. In fact, certain CIDNP observations may depend on rapid quenching of spin-polarized triplets [162)].

In this regard, the interpretation of CIDNP observations during irradiation of ketone-amine solutions is interesting. Irradiation of acetonitrile solutions of several benzophenones produces spin-polarization of the benzophenone protons if 1,4-diazabicyclo[2.2.2]octane (DABCO) is present [163)]. The polarizations were interpreted as arising from radical ions. Since DABCO has an oxidation potential even lower than that of dialkylanilines, Mataga's work [157)] suggests that DABCO can indeed transfer an electron to triplet ketones. Interestingly, quantum yields for ketone photoreduction by DABCO are not as large as for amines which apparently do not yield radical ions [152)].

2. n,π^* vs π,π^* Reactivity

Ketones with π,π^* lowest triplets are photoreduced by tertiary amines, but quantum efficiencies vary with solvent polarity. For fluorenone, photoreduction quantum yields decrease with increasing solvent polarity [164)]; for naphthyl ketones, the dependence is reversed [168)]. In both cases, rate constants for quenching of triplet ketone by amine are much higher in polar (both protic and aprotic) than in nonpolar solvents. This fact suggests a high degree of charge transfer in the interaction. In fact, Davidson has observed the radical ions from fluorenone in polar solvents [166)]. In contrast, ketones with n,π^* lowest triplets are quenched by tertiary amines no faster in acetonitrile than in hydrocarbon solvents and appreciably more slowly in protic solvents [152,155)]. Hydrogen bonding to the amine lone pair can explain the retardation of CT quenching of n,π^* triplets in protic solvents. The absence of this solvent effect in naphthyl ketones suggests that amines quench n,π^* and π,π^* triplets by somewhat different mechanisms.

There is really no solid evidence regarding the relative reactivities of n,π^* and π,π^* triplets in CT quenching. The rate constants for tertiary amine quenching of triplet naphthyl ketones are in the 10^5—10^6 M^{-1} sec^{-1} range, as compared to values greater than 10^9 M^{-1} sec^{-1} for quenching of triplet benzophenone, acetophenone, etc. A good deal of this difference must reflect the low triplet excitation energy of naphthyl ketones. Although rates of CT quenching have been reported for several different ketones with varying triplet energies, reduction potentials, and electronic configuration, variations in donor structure and in solvent polarity are such that no extensive correlation is possible.

3. Aminoketones

Intramolecular CT quenching is interesting for several reasons. Since bimolecular quenching of many triplet ketones by tertiary amines is so rapid as to approach being diffusion-controlled, analogous intramolecular quenching should be rotation-controlled [167]. That is, the rate limiting step may involve those combinations of bond rotations which bring the molecule into one of several conformations in which the nitrogen lone pair overlaps significantly with the carbonyl molecular orbitals. One would think that, in acyclic molecules, the fewer the carbons between the two functional groups, the faster an unfavorable conformation can rotate into a favorable one and the greater the proportion of favorable conformers at any given moment.

We have studied several aminoketones and the results are not totally understandable [168]. For α,γ, and δ dialkylaminosubstituted ketones, the quantum yield for formation of long-lived triplet increases as the amino group gets farther away from the carbonyl and is higher in protic solvents than in aprotic solvents.

$$\mathrm{Ph\underset{\displaystyle C}{\underset{\|}{O}}(CH_2)_nNR_2}$$

This trend can be explained in either of two ways. First, rates of singlet state CT quenching may be so great that they compete with intersystem crossing — and compete better the closer together the two functional groups are. The protic solvent effect would then be the expected retardation of CT because of hydrogen bonding to the lone pair. Second, large fractions of the ground state molecules may exist in favorable conformations such that CT is effectively the only reaction following intersystem crossing. The measured triplet yields then would define not the competition between intersystem crossing and singlet reaction but rather relative conformational populations.

The actual rate constants for intramolecular quenching in the triplet aminoketones are $<10^9$, 7×10^9, 3×10^9 sec^{-1} for $n = 1,3,4$, respectively [168]. The rates for the γ- and δ-aminoketones are in the range expected for bond rotations; that for the α-aminoketone is surprisingly low. In fact, rate constants for singlet CT quenching would have to be at least ten times faster than the triplet rates if the low measured triplet yields are due to singlet reaction.

That close overlap of the nitrogen lone pair with carbonyl orbitals is required for CT quenching is demonstrated by the behavior of the 4-benzoylpiperidine below [169]. It forms two kinetically distinct triplets, one of which undergoes, at a rate $>10^9$ sec^{-1}, solely type II cyclization; the other of which undergoes, at a rate of 10^7 sec^{-1}, only α-cleavage. The two conformers with axial and equatorial benzoyl groups, respectively, seem to be the likely triplets. The latter triplet,

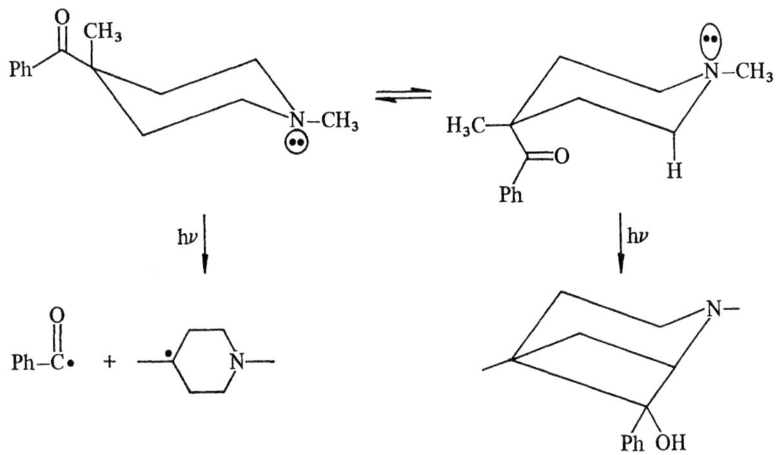

in which the nitrogen lone pair is held some 6Å from the carbonyl, undergoes CT quenching at least three orders of magnitude more slowly than do acyclic γ-dialkylamino phenyl ketones.

γ-amino ketones provide one further interesting bit of information about the complexes formed from triplet carbonyls and amines. The γ-dimethylamino naphthyl ketone below undergoes an unquenchable type II photoelimination with low efficiency (<1%) in all solvents. No quenchable triplet reaction occurs in aprotic solvents, although long-lived (π,π*) triplets are formed with high efficiency [170]. No γ-hydrogen abstraction occurs and the expected CT quenching does not produce any diradical. In protic solvents, however, the ketone undergoes

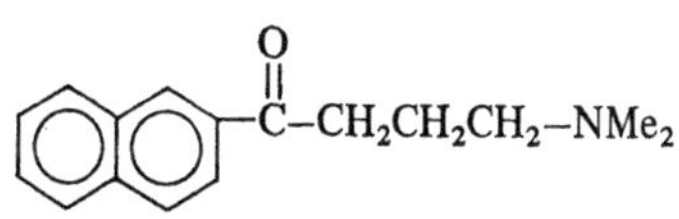

an efficient, readily quenchable triplet state photoelimination. A protic solvent is required for the cyclic excited CT complex to rearrange. Presumably the cyclic nature of the complex holds the γ-protons away from the carbonyl oxygen, such that the facile proton transfer which occurs in bimolecular complexes cannot occur.

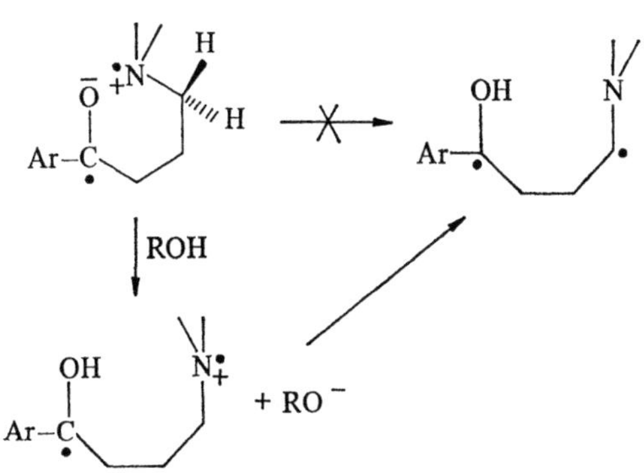

D. Generality of CT Quenching

Weller [171] has pointed out that the rate constants for electron transfer to fluorescent aromatics are diffusion-controlled when the process is exothermic but are proportional to the free energy change when the process is endothermic. Here $E(D/D^+)$ is the donor oxidation potential, positive by convention, and $E(A^-/A)$

$$\Delta G = -E_T + E(D/D^+) - E(A^-/A) - T\Delta S - \frac{e^2}{\varepsilon r} \tag{3}$$

is the acceptor reduction potential, negative by convention. These can be measured polarographically relative, of course, to the same reference electrode. The last term describes the Coulombic attraction of two oppositely charged ions. It must be stressed that this equation pertains to complete electron transfer with formation of a caged pair of radical ions. Solvent effects on the rates of such processes are dramatic. As mentioned above, solvent effects in CT quenching of triplet ketones are minimal and radical ions are rarely formed.

The thermodynamics of electron transfer can influence CT quenching only inasmuch as electron transfer contributes to the excited CT complex(exciplex). In valence bond terms, the value of lc^2 describes the percentage of electron

$$A^* + D \longrightarrow \dot{A}^- \dot{D}^+$$

$$\psi_{CT} = a\phi(A^* + D) + lc\phi(\dot{A}^- + \dot{D}^+) \tag{4}$$

transfer in the complex. Cohen [172] and we [130] pointed out that the rate constants describing quenching of $^3n,\pi^*$ ketones by amines, sulfides, and olefins correlate with the ionization potentials of the donors, as might be expected for a CT process [173]. There are three specific aspects of such correlations which demand comment.

First, although ionization potentials are proportional to oxidation potentials, both effectively measure unimolecular processes. Therefore, any steric effects which may hinder the formation of a bimolecular CT complex are not accounted for by simple redox potentials. It is clear that steric factors do influence rates of CT quenching [130,133,174].

Second, plots of log k_q of a given triplet ketone *vs.* ionization potential or oxidation potential are roughly linear for the donors listed above. The slopes are negative and have values which indicate that variations in rates respond to only 15—20% the variations in endothermicities of full electron transfer. It has been suggested [172] that one might infer from this correlation that lc^2 is 0.15—0.20 in triplet exciplexes of n,π^*-excited ketones.

Fewer data exist for constant donor and varying ketone, but presumably there should be a correlation between log k_q and $[\Delta E_T + E(A^-/A)]$. With ketones and amines, what results exist are inconclusive [175].

Third, does any compound with a given ionization or oxidation potential quench a given triplet ketone at the same rate? The answer is no for two reasons. Hydrogen abstraction and electronic energy transfer are discrete, competitive bimolecular quenching processes. Even in the absence of one of these competing processes, CT quenching rates depend on the compound type as well as on its oxidation potential [130,176]. In particular, benzene and some of its simple derivatives quench triplet ketones by a process which has been inferred to involve charge transfer because of linear log *k-IP* correlations [132,172]. Presumably, the aromatics act as π-donors. However, the linear free energy plot which correlates their quenching ability lies some 2 units (log 100) below the plot for n-donors and unhindered olefins. In other words, the latter donors quench some 100 times faster than do benzenes. There is no adequate explanation for this difference — nor for why n-donors and olefins should be correlated by the same parameters. It was not too many years ago that it was discovered that the relatively short lifetimes of triplet ketones in benzene was not intrinsic but was due to quenching by the solvent [177]. It is still not clear what combination of radical-like addition, CT complex formation, and as yet undefined processes contribute to the quenching of triplet ketones by benzene solvent [178,179]. This topic will be reexamined below.

It is interesting to compare how CT quenching might compete with the other bimolecular reactions of triplet ketones. For example, tetrabutylstannane and dibutylsulfide have comparable gas phase ionization potentials. The latter, having a readily available lone pair of electrons, quenches triplet ketones rapidly, whereas the latter does so much more slowly and primarily as a hydrogen atom donor [180]. Conjugated dienes have ionization potentials ~8.5 eV. If the behavior of monoolefins can be used as a guide, the conjugated dienes should quench triplet ketones by CT interactions with rate constants in the range 10^8–10^9 M^{-1} sec^{-1} [130]). In fact, conjugated dienes quench triplet ketones by energy transfer at rates approaching those of diffusion. However, triplet ketones do add to dienes to produce vinyloxetanes, but in very low quantum efficiency [181]. Apparently a few percent CT quenching does compete with energy transfer. We have already [130] pointed out that with 1,3-pentadiene, which is used widely to monitor triplet formation [5], CT quenching would be predicted to contribute at most 1% to its total rate of quenching, at least for ketones such as benzophenone and acetophenone, in which $E_T + E(A^-/A)$ is ~28 kcal/mole.

As mentioned above, there is a scarcity of data on how variations in ketone structure affect CT quenching rates. A comparison of the interactions of substituted benzenes with triplet acetophenone and triplet α-trifluoroacetophenone is interesting [132,182]. The relative reactivities of various hydrocarbons towards triplet acetophenone are suggestive of direct hydrogen atom abstraction by the triplet ketone. The low reduction potential of the trifluroketone enhances the rate of CT quenching so much that photoreduction proceeds almost entirely by a

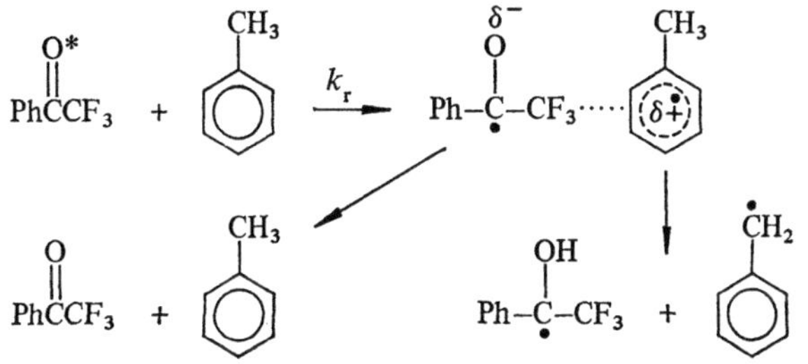

CT interaction [132]. In the case of substituted benzenes with benzylic hydrogens, proton transfer yields radicals. The main evidence for this CT photoreduction mechanism is the lack of any isotope effect on k_r when toluene-α-d_3 is substituted for toluene-h_8. In the absence of any labile protons on the aromatic half of the CT complex, such as with benzene itself, the CT interaction is primarily a quenching one, unless an external proton source is present, in which case the complex is apparently protonated [183].

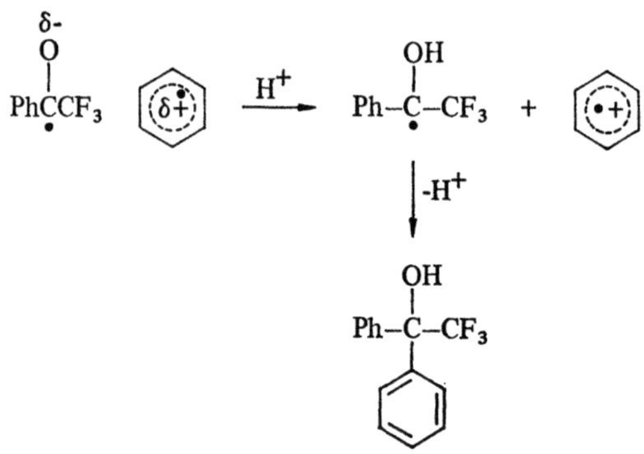

The rates at which several substituted benzenes quench triplet benzophenone have been measured [178]. No single linear free energy relationship can be derived. For alkoxybenzenes, alkylbenzenes, benzene itself, and benzotrifluoride as quenchers, one finds a linear plot of log k_q *vs.* *IP* with a slope similar to that found for the plot of all substituted benzenes and triplet α-trifluoroacetophenone. A given aromatic such as benzene quenches the fluorinated ketone triplet, which has an $E_T + E(A^-/A)$ value of only 16 kcal, some 50 times faster than it quenches triplet acetophenone or benzophenone [132]. This rate difference reflects only ~20% of the full 12 kcal difference in thermodynamic redox potentials. However, the halobenzenes and benzonitrile quench triplet benzophenone faster than does benzene [178]. It seems likely that with these electron poor benzene derivatives, some alternate chemical reaction becomes dominant. Although a reverse CT process has been suggested, with the triplet ketone as *donor*, it is perhaps more likely that some sort of radical addition occurs with conjugating substituents on

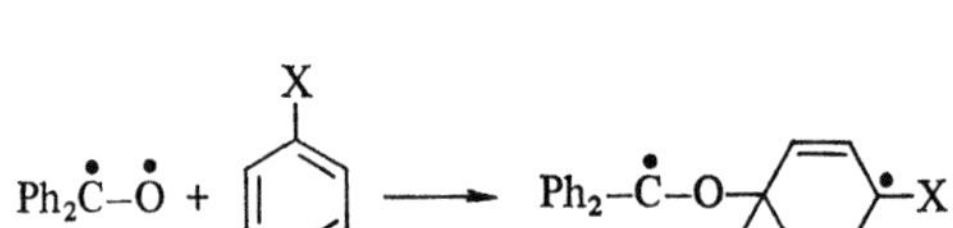

the benzene ring being able to enhance the process somewhat. This interpretation would perhaps explain the low quenching rate of benzotrifluoride. The matter is by no means settled, but it may be relevant to note here that trialkylboranes also readily quench triplet benzophenone by a process which certainly resembles the addition of oxygen radicals to boranes [184].

$$Ph_2\dot{C}{-}\dot{O} + R_3B \longrightarrow Ph{-}\dot{C}{-}\overset{\cdot +}{O}{-}\overset{-}{B}R_3 \longrightarrow Ph{-}\dot{C}{-}OBR_2 + R\cdot$$

As indicated above, it is unlikely that free radical ions are formed in these CT quenching processes. If they were, protons more acidic than benzylic ones could be lost. For example, the radical cation of phenoxyacetic acid readily decarbonylates. Thus, the report that benzophenone can photosensitize the

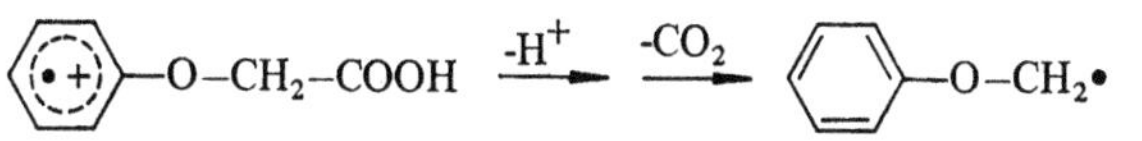

decarboxylation of several aryl acids [185] is quite interesting. This observation does not necessarily indicate the presence of any free radical cations, since a cyclic proton transfer process could occur in the excited CT complex. It remains

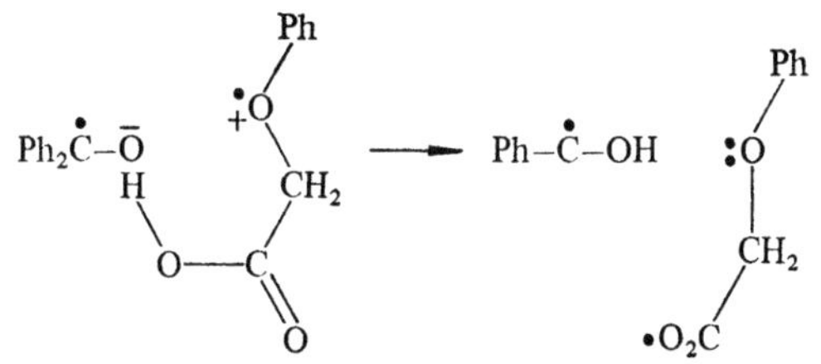

possible that the photoreduction of simple ketones such as benzophenone and acetophenone by toluene involves competing CT and direct hydrogen abstraction.

E. Self-Quenching

In the past few years it has become apparent that triplet phenyl ketones are quenched by ground state ketone molecules. In the case of Michler's ketone, this observation [186] should not have been surprising, since Michler's ketone is a dimethylaniline and dimethylanilines quench even π,π^* ketone triplets very rapidly [187]. With benzophenone, one might argue that the ketone is merely a substituted benzene. However, ground state benzophenone quenches triplet benzophenone with a rate constant on the order of 10^6 M^{1-} sec^{-1} [188], approximately the same as that recorded for benzonitrile [178] and some 100 times larger than that for benzene itself. Clearly, the self-quenching is not a case of charge transfer from ground state to excited triplet. That benzonitrile and benzophenone are comparable quenchers of triplet benzophenone rules out the possibility that excimer formation accounts for usually rapid self-quenching. As discussed above, it remains to be determined by what mechanism electron-deficient benzenes quench triplet ketones.

VI. Eliminations of α-Substituents

As mentioned above, a variety of α-substituted ketones undergo light-induced cleavage of that substituent. The simple picture presented earlier for an n,π^* excited state suggests that both ionic and radical cleavage is possible and both have in fact been observed [17,189]. The simple model would suggest that ionic cleavage occurs only from excited singlets, whereas radical cleavage could also be a triplet reaction. The experimental evidence bearing on this hypothesis

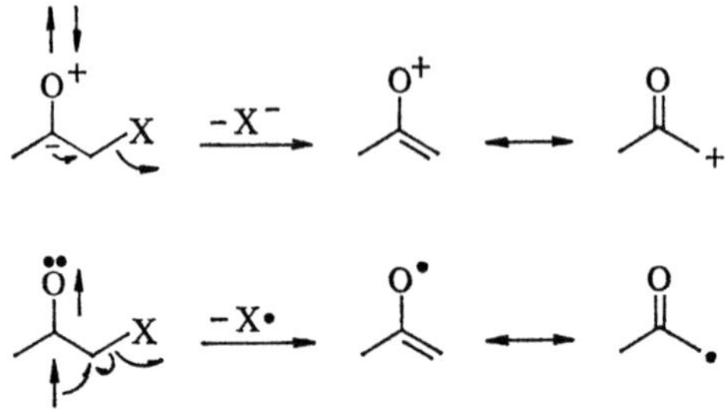

is confusing. In fact, there really is not much quantitative information on rates of excited state processes in α-substituted ketones.

The situation is further complicated by the well known fact that α-substituents affect the $n \rightarrow \pi^*$ transitions [190], so that it is tempting to speculate upon how electronic interaction of the α-substituent with the carbonyl may predispose the excited state toward fragmentation. For example, it is well known that, in cyclic ketones, axial α-substituents produce the largest changes in $n \rightarrow \pi^*$ transitions [191]. Likewise, amino [192] and tosyloxy [193] groups are photocleaved when axial but not when equatorial. Presumably the α-C-X σ-orbital must be parallel to the carbonyl π-system.

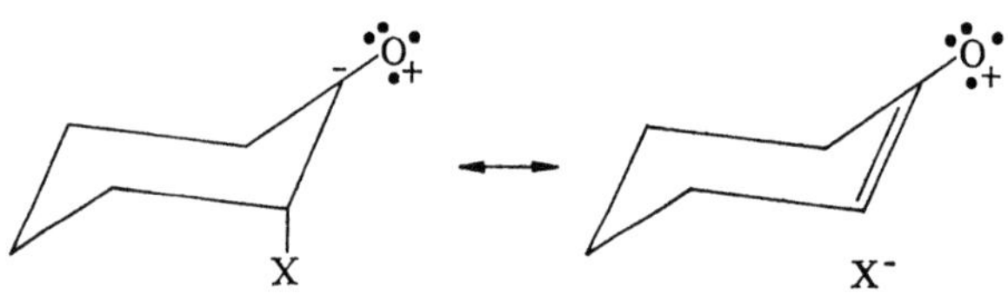

Another example of such complications are the differing photoreactions of *para*'- and *meta*'-methoxy benzoin acetates [194]. The *para*-isomer undergoes predominately α-cleavage and a little cyclization, both apparently from the triplet.

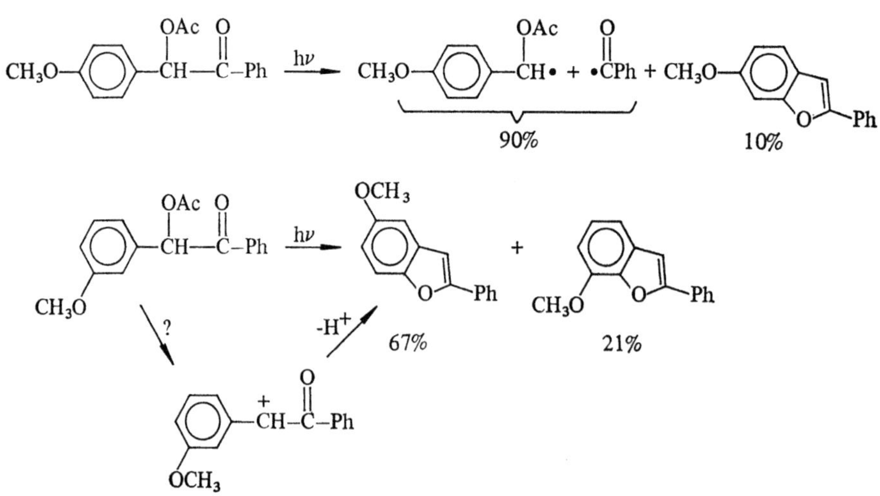

The *meta*-isomer eliminates acetic acid and cyclizes, apparently from a singlet excited state. This cyclization can certainly be visualized to involve ionic cleavage, although the original investigators suggested instead an "ephemeral" tricyclic oxetane intermediate. The uv spectra of α-arylketones indicate that there is considerable interaction between the aryl group and the carbonyl in the n,π^* states [195]. It is well known that *meta*-methoxybenzyl acetates photosolvolyze far better than the *para*-isomers [196]. Inasmuch as delocalization of excitation onto the α-aryl group occurs, the tendency toward elimination of the acetate ion would be enhanced over that already present in the carbonyl n,π^* singlet. The *p*-methoxybenzoin acetate does not undergo this rapid singlet reaction and undergoes normal triplet state chemistry. The mechanism for the small amount of competing cleavage cannot be readily deduced.

Not enough is known for one to predict whether ionic or radical cleavage will occur. Many α-chloro and α-bromo phenyl ketones are used as photoinitiators for polymerizations [52], so they clearly produce radicals readily. Irradiation of chloroacetone in solution initiates the addition of CCl_4 and thiols to olefins [197]. Careful analysis of product structures suggests that only radical cleavage occurs. For example, in anisole the main product is *ortho*-methoxyphenylacetone. Radicals but not carbonium ions add preferentially *ortho* to monosubstituted benzenes.

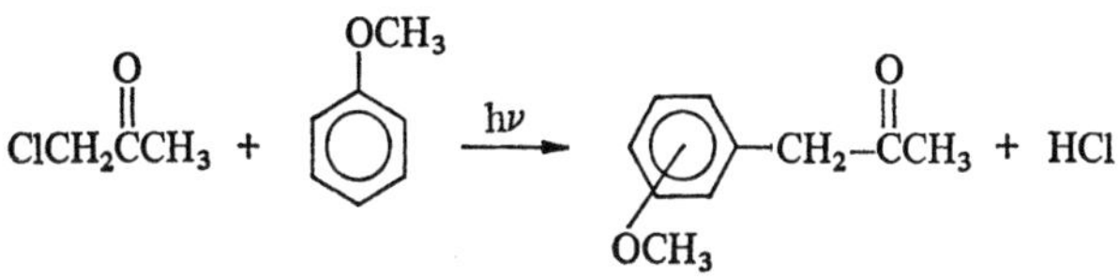

o:*m*:*p*=85:7:5

Rearrangements characteristic of carbonium ions have been cited as evidence for ionic cleavages [193].

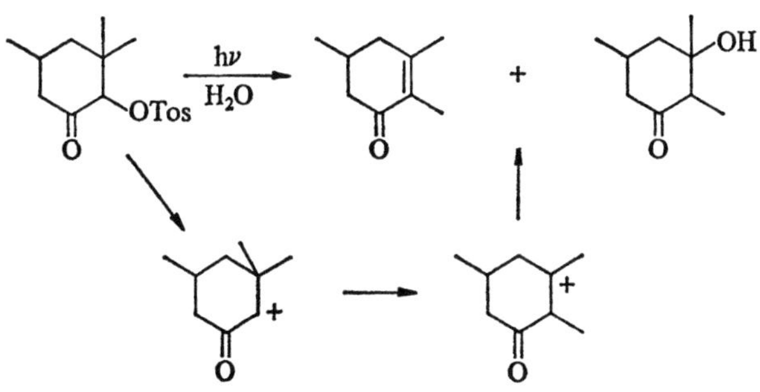

We have looked at the photochemistry of α-chloro-α-methylvalerophenone and find it to be almost hopelessly complex [198]. At high conversion, the major photoproducts are the β,γ-unsaturated ketones plus β and γ-chloroketones formed by addition of HCl to them. However, the reaction is quite complex, since many other products are formed only at low conversion. All of these, including some but not all of the β,γ-unsaturated ketones, are quenched by radical traps and triplet

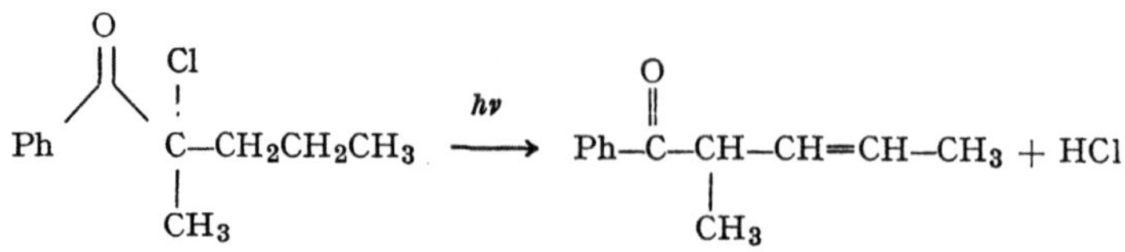

quenchers. Apparently a very short-lived (~0.2 nsec) triplet undergoes primarily cleavage of atomic chlorine to give a host of radical products, including the β,γ-unsaturated ketone. Since the amount of unquenchable β,γ-unsaturated formation is largest in polar solvents, some ionic singlet cleavage is likely.

On the other hand, Schaffner has reported quenching and sensitization experiments on 3-mesyloxy-2-butanone which suggest radical cleavage from the singlet and ionic cleavage from the triplet [189]! Despite the wealth of knowledge on ketone photochemistry, the details of the various β-cleavage reactions remain mysterious.

VII. Summary and Sermon

Structure-reactivity relationships are now well understood for hydrogen abstraction and α-cleavage reactions of monofunctional excited ketones. The generality of CT quenching is recognized but many aspects are poorly understood. Most aspects of β-cleavage reactions are poorly understood.

When one takes up a polyfunctional molecule which contains at least one carbonyl group, one knows which reactions of that carbonyl group are possible. One does not know how well these reactions will compete with various physical and chemical interactions of that excited carbonyl group with other functional groups in the molecule. It is this aspect of carbonyl photochemistry which requires and deserves extensive future research. The uniquely well understood chemistry of the carbonyl group can serve as a monitor for studying interactions in electronically excited polyfunctional molecules.

VIII. References

1) Photochemistry, Vol. 4. A specialist periodical report. London: The Chemical Society 1973.
2) a) Zimmermann, H. E.: Advan. Photochem. *1*, 183 (1963);
b) Wagner, P. J., Hammond, G. S.: Advan. Photochem. *5*, 168 (1968);
c) Dalton, J. C., Turro, N. J.: Ann. Rev. Phys. Chem. *21*, 499 (1970);
d) Kossanyi, J., Furth, B.: L'Actualité Chim. *2*, 7; *3*, 3 (1974).
3) Wagner, P. J.: In: Creation and detection of the excited state (A. A. Lamola, ed.); Vol. 1A, p. 173. New York: Marcel Dekker 1971.
4) Ishikawa, H., Noyes, W. A. Jr.: J. Am. Chem. Soc. *84*, 1502 (1962).
5) a) Lamola, A. A., Hammond, G. S.: J. Chem. Phys. *43*, 2129 (1965);
b) Cundall, R. B., Griffiths, P. A.: Trans. Faraday Soc. *61*, 1968 (1965).
6) a) Lamola, A. A.: In: Technique of organic chemistry (P. A. Leermakers and A. Weissberger, eds.), Vol. XIV, p. 17. Interscience 1969.
b) Wagner, P. J., Kochevar, I.: J. Am. Chem. Soc. *90*, 2232 (1968).
7) Wettack, F. S., Renkes, G. D., Rockley, M. G., Turro, N. J., Dalton, J. C.: J. Am. Chem. Soc. *92*, 1793 (1970).
8) Wagner, P. J.: Mol. Photochem. *3*, 169 (1971).
9) Wagner, P. J.: Accounts Chem. Res. *4*, 168 (1971).
10) Robinson, G. W., DiGiorgio, V. E.: Can. J. Chem. *36*, 31 (1958).
11) Wagner, P. J., May, M. J., Haug, A.: Chem. Phys. Letters *13*, 545 (1972).
12) Saltiel, J., Curtis, H. C., Metts, L., Miley, J. W., Winterle, J., Wrighton, M.: J. Am. Chem. Soc. *92*, 410 (1970).
13) Marsh, G., Kearns, D. R., Schaffner, K.: Helv. Chim. Acta. *51*, 1890 (1968).
14) a) Lamola, A. A.: J. Chem. Phys. *47*, 4810 (1967);
b) Li, Y. H., Lim, E. C.: Chem. Phys. Letters *7*, 15 (1970).
15) Goodman, L., Koyanagi, M.: Mol. Photochem. *4*, 365 (1972).
16) Walling, C., Wagner, P. J.: J. Am. Chem. Soc. *86*, 3368 (1964).
17) Schaffner, K.: Pure Appl. Chem. *16*, 75 (1967).
18) Noyes, W. A., Jr., Porter, G. B., Jolley, J. E.: Chem. Rev. *56*, 49 (1956).
19) Doepker, R., Mains, G. J.: J. Phys. Chem. *66*, 690 (1962).
20) Vaish, S. P., McAlpine, R. D., Cocivera, M.: J. Am. Chem. Soc. *96*, 1683 (1974).
21) Wagner, P. J.: J. Am. Chem. Soc. *88*, 5672 (1966).
22) Jarvie, J. M., Laufer, A. H.: J. Phys. Chem. *68*, 2557 (1964).
23) Bibart, C. H., Rockley, M. G., Wettack, F. S.: J. Am. Chem. Soc. *91*, 2802 (1969).
24) a) Den-Hollander, J. A., Kaptein, R., Brand, P. A.: Chem. Phys. Letters *10*, 430 (1971);
b) Tomkiewicz, M., Groen, A., Cocivera, M.: J. Chem. Phys. *56*, 5850 (1972).
25) Yang, N. C., Feit, E. D., Hui, M. H., Turro, N. J., Dalton, J. C.: J. Am. Chem. Soc. *92*, 6974 (1970).
26) a) Engel, P. S.: J. Am. Chem. Soc. *92*, 6074 (1970);
b) Robbins, W. K., Eastman, R. H.: J. Am. Chem. Soc. *92*, 6077 (1970).
27) Walling, C., Clark, R. T.: J. Am. Chem. Soc. *96*, 4530 (1974).
28) Dalton, J. C., Dawes, K., Turro, N. J., Weiss, D. S., Barltrop, J. A., Coyle, J. D.: J. Am. Chem. Soc. *93*, 7213 (1971).
29) Salem, L.: J. Am. Chem. Soc. *96*, 3486 (1974).
30) a) Hayes, D. M., Morokuma, K.: Chem. Phys. Letters *12*, 539 (1972);
b) Fink, W. H.: J. Am. Chem. Soc. *94*, 1073, 1078 (1972).
31) Solomon, J., Jonah, C., Chandra, P., Bersohn, R.: J. Chem. Phys. *55*, 1908 (1971).
32) Baggiolini, E., Hamilow, H. P., Schaffner, K.: J. Am. Chem. Soc. *92*, 4906 (1970).
33) Closs, G. L., Doubleday, C. E.: J. Am. Chem. Soc. *94*, 9248 (1972).
34) Thomas, M., Wagner, P. J.: unpublished results.
35) Kaptein, R., Freeman, R., Hill, H. D. W.: Chem. Phys. Letters *26*, 104 (1974).
36) a) Butenandt, A., Paschmann, L.: Ber. *77*, 394 (1944);
b) Wehrli, H., Schaffner, K.: Helv. Chim. Acta., *45*, 385 (1962);
c) Quinkert, G., Heine, H. G., Tetrahedron Letters, 1659 (1963).
37) Barltrop, J. A., Coyle, J. D.: Chem. Commun. *1969*, 1081.

38) Wagner, P. J., Spoerke, R. W.: J. Am. Chem. Soc. *91*, 4437 (1969).
39) Dalton, J. C., Pond, D. M., Weiss, D. S., Lewis, F. D., Turro, N. J.: J. Am. Chem. Soc. *92*, 2564 (1970).
40) a) Beard, J. M., Eastman, R. H.: Tetrahedron Letters *1970*, 3029;
b) Yang, N. C., Chen, R. H.-K.: J. Am. Chem. Soc. *94*, 530 (1972).
41) a) Yates, P., Loutfy, R. O.: Accounts Chem. Res. *6*, 209 (1975);
b) Yates, P.: Pure Appl. Chem. *16*, 93 (1968).
42) Collins, P. M., Oparaeche, N. N., Whitton, B. R.: J. Chem. Soc., Chem. Commun. *1974*, 292.
43) Morton, D. R., Turro, N. J.: J. Am. Chem. Soc. *95*, 3947 (1973).
44) Lee, E. K. C., Hemminger, J. C., Rubutt, C. F.: J. Am. Chem. Soc. *93*, 1867 (1971).
45) a) Hostettler, H. U.: Helv. Chim. Acta *49*, 2417 (1966).
b) Morton, D. R., Lee-Ruff, E., Southam, R. M., Turro, N. J.: J. Am. Chem. Soc. *92*, 4349 (1970).
46) a) Quinkert, G., Jacobs, P., Stohrer, W.-D.: Angew. Chem. Intern. Ed. Engl. *13*, 197 (1974);
b) Stohrer, W.-D.: Jacobs, P., Kaiser, K. H., Wiech, G., Quinkert, G.: Topics Curr. Chem., *46*, 181 (1974).
47) a) Wagner, P. J., McGrath, J. M.: J. Amer. Chem. Soc., *94*, 3849 (1972);
b) Lewis, F. D., Hilliard, T. A.: J. Am. Chem. Soc. *94*, 3852 (1972).
48) Heine, H.-G., Hartmann, W., Kory, D. R., Magyar, J. G., Hoyle, C. E., McVey, J. K., Lewis, F. D.: J. Org. Chem. *39*, 691 (1974).
49) Bartlett, P. D., Hiatt, R. R.: J. Am. Chem. Soc. *80*, 1398 (1958).
50) Lewis, F. D., Magyar, J. G.: J. Am. Chem. Soc. *95*, 5973 (1973).
51) Zepp, R. G., Wagner, P. J.: Chem. Commun. *1972*, 167.
52) Heine, H.-G., Rosenkranz, H.-J., Rudolph, H.: Angew. Chem. Intern. Ed. Eng. *11*, 974 (1972).
53) Lewis, F. D., Magyar, J. G.: J. Org. Chem. *37*, 2102 (1972).
54) Hammond, G. S., Leermakers, P. A.: J. Am. Chem. Soc. *84*, 207 (1962).
55) Yang, N. C., McClure, D. S., Murov, S. L., Houser, J. J., Dusenbery, R., J.: Am. Chem. Soc. *89*, 5466 (1967).
56) Wagner, P. J., Kemppainen, A. E.: J. Am. Chem. Soc. *90*, 5898 (1968).
57) Yang, N. C., Dusenbery, R.: J. Am. Chem. Soc. *90*, 5899 (1968).
58) Wagner, P. J., Kemppainen, A. E., Schott, H. N.: J. Am. Chem. Soc. *95*, 5604 (1973).
59) Wagner, P. J., Schott, H. N.: J. Am. Chem. Soc. *91*, 5383 (1969).
60) DeBoer, C. D., Herkstroeter, W. G., Marchetti, A. P., Schultz, A. G., Schlessinger, R. H.: J. Am. Chem. Soc. *95*, 3963 (1973).
61) a) Herz, W., Nair, M. G.: J. Am. Chem. Soc. *89*, 5475 (1967);
b) Kropp, P. J.: J. Am. Chem. Soc. *91*, 5783 (1969);
c) Rosenberg, H. M., Serve, P.: J. Am. Chem. Soc. *95*, 4746 (1970);
d) Hornback, J. M.: J. Am. Chem. Soc. *96*, 6773 (1974).
62) a) Reinfried, R., Bellus, D., Shaffner, K.: Helv. Chim. Acta *54*, 1517 (1971).
b) Smith, A. B., Agosta, W. C.: J. Am. Chem. Soc. *95*, 1961 (1973).
63) Harris, E. G.: M. S. Thesis, Michigan State University, 1973.
64) Giering, L., Berger, M., Steel, C.: J. Am. Chem. Soc. *96*, 953 (1974).
65) Lutz, H., Duval, M. C., Breheret, E., Lindqvist, L.: J. Phys. Chem. *76*, 821 (1972).
66) Mallory, F. B., Arnold, M. T., Mallory, C. W., Colman, T. E.: Abstracts, 5th Northeast Regional Meeting, American Chemical Society, Rochester, N. Y., October, 1973, No. 26.
67) Wagner, P. J., Kemppainen, A. E.: J. Am. Chem. Soc. *94*, 7495 (1972).
68) Lewis, F. D., Johnson, R. W., Kory, D. R.: J. Am. Chem. Soc. *96*, 6100 (1974).
69) Previtali, C. M., Scaiano, J. C.: J. Chem. Soc., Perkin Trans. *2*, 1667, 1672 (1972).
70) Evans, M. G., Polanyi, M.: Trans. Faraday Soc. *34*, 11 (1938).
71) Hochstrasser, R. M., Liu, T.-S.: J. Chem. Phys. *49*, 4929 (1968).
a) Turro, N. J., Lee, T.: J. Am. Chem. Soc. *91*, 5651 (1969).
72) Wagner, P. J., Kelso, P. A., Kemppainen, A. E., Zepp, R. G.: J. Am. Chem. Soc. *94*, 7500 (1972).
73) Lewis, F. D., Johnson, R. W., Ruden, R. A.: J. Am. Chem. Soc. *94*, 4292 (1972).

74) Wagner, P. J., Zepp, R. G.: J. Am. Chem. Soc. *93*, 4958 (1971).
75) Walling, C., Padwa, A.: J. Am. Chem. Soc. *85*, 1597 (1963).
76) Turro, N. J., Weiss, D. S.: J. Am. Chem. Soc. *90*, 2185 (1968).
77) a) Jeger, O., Schaffner, K.: Pure Appl. Chem. *21*, 247 (1970);
b) Sugiyama, N., Yamada, K., Hoyama, H.: Chem. Commun. *1968*, 1254.
78) Sugiyama, N., Nishio, T., Yamada, K., Aoyama, H.: Bull. Chem. Soc. Japan *43*, 1879 (1970).
79) Breslow, R., Winnik, M.: J. Am. Chem. Soc. *91*, 3083 (1969).
80) Winnik, M. A., Trueman, R. E., Jackowski, G., Saunders, D. S., Whittington, S. G.: J. Am. Chem. Soc. *96*, 4843 (1974).
81) Winnik, M. A., Lee, C. K., Basu, S., Saunders, D. S.: J. Am. Chem. Soc. *96*, 6182 (1974).
82) Breslow, R., Baldwin, S. W.: J. Am. Chem. Soc. *92*, 732 (1970).
83) Padwa, A., Alexander, E., Niemcyzk, M.: J. Am. Chem. Soc. *91*, 465 (1969).
84) Alexander, E. C., Uliana, J. A.: J. Am. Chem. Soc. *96*, 5644 (1974).
85) Padwa, A., Eisenberg, W.: J. Am. Chem. Soc. *94*, 5859 (1972).
86) Lewis, F. D., Johnson, R. W., Johnson, D. E.: J. Am. Chem. Soc. *96*, 6090 (1974).
87) a) Zwicker, E. F., Grossweiner, L. I., Yang, N. C.: J. Am. Chem. Soc. *85*, 2671 (1963);
b) Huffman, K. R., Loy, M., Ullmann, E. F.: J. Am. Chem. Soc. *87*, 5417 (1965);
c) Porter, G., Tchir, M. F.: J. Chem. Soc. (A) *1971*, 3772.
88) Wagner, P. J., Chen, C.-P.: J. Am. Chem. Soc. *98*, 239 (1976).
89) Dalton, J. C., Chan, H.-F.: J. Am. Chem. Soc. *95*, 4085 (1973).
90) Yang, N. C., Elliot, S. P.: J. Am. Chem. Soc. *91*, 7550 (1969). — Yang, N. C. Elliot, S. P., Kim, B.: J. Am. Chem. Soc. *91*, 7551 (1969).
91) a) Stephenson, L. M., Cavigli, P. R., Parlett, J. L.: J. Am. Chem. Soc. *93*, 1984 (1971);
b) Casey, C. P., Boggs, R. A.: J. Am. Chem. Soc. *94*, 6457 (1972).
92) Heller, A.: Mol. Photochem. *1*, 257 (1969).
93) Hammond, G. S.: Advan. Photochem. *7*, 373 (1969).
94) Michl, J.: Mol. Photochem., *4*, 257 (1972); Topics Curr. Chem. *46*, 1 (1974).
95) Salem, L., Leforestier, C., Segal, G., Wetmore, R.: J. Am. Chem. Soc. *97*, 479 (1975).
96) Wagner, P. J., Kelso, P. A., Zepp, R. G.: J. Am. Chem. Soc. *94*, 7480 (1972).
97) Dauben, W. G., Salem, L., Turro, N. J.: Accounts Chem. Res. *8*, 41 (1975).
98) Charney, D. R., Dalton, J. C., Hautala, R. R., Snyder, J. J., Turro, N. J.: J. Am. Chem. Soc. *96*, 1407 (1974).
99) a) Wagner, P. J.: J. Am. Chem. Soc. *89*, 2503 (1967);
b) Wagner, P. J.: Tetrahedron Letters *1968*, 5385.
100) Keller, P., Eggart, G., Wehrli, H., Schaffner, K., Jeger, O.: Helv. Chim. Acta *50*, 2259 (1967).
101) Wagner, P. J., Stratton, T. S.: unpublished results. — Turro, N. J., Dalton, J. C.: private communication.
102) Wagner, P. J., Meuller, W.: unpublished results.
103) Bartlett, P. D., Porter, N. A.: J. Am. Chem. Soc. *90*, 5317 (1968).
104) Wagner, P. J., Zepp, R. G.: J. Am. Chem. Soc. *94*, 287 (1972).
105) Wagner, P. J., Liu, K.-C.: J. Am. Chem. Soc. *96*, 5952 (1974).
106) Hoffman, R., Swaminathau, S., Odell, B. G., Gleiter, R.: J. Am. Chem. Soc. *92*, 7081 (1970).
107) Stephenson, L. M., Braumann, J. I.: J. Am. Chem. Soc. *93*, 1988 (1971).
108) Wagner, P. J., Kelso, P. A., Kemppainen, A. E., McGrath, J. M., Schott, H. N., Zepp, R. G.: J. Am. Chem. Soc. *94*, 7506 (1972).
109) Orban, I., Schaffner, K., Jeger, O.: J. Am. Chem. Soc. *85*, 3033 (1963).
110) Turro, N. J., Lee, T. J.: J. Am. Chem. Soc. *92*, 7467 (1970).
111) Stephenson, L. M., Gibson, T. A.: J. Am. Chem. Soc. *94*, 4599 (1972).
112) Wagner, P. J., Kemppainen, A. E.: J. Am. Chem. Soc. *90*, 5896 (1968).
113) Gagosian, R. B., Dalton, J. C., Turro, N. J.: J. Am. Chem. Soc. *92*, 4752 (1970).
114) a) Pryor, W. A., Davis, W. H., Jr., Stanley, J. P.: J. Am. Chem. Soc. *95*, 4754(1973);
b) Henderson, R. W.: J. Am. Chem. Soc. *97*, 213(1975).
115) a) Wagner, P. J.: Tetrahedron Letters *1967*, 1753;
b) Wagner, P. J., Kochevar, I., Kemppainen, A. E.: J. Am. Chem. Soc. *94*, 7489 (1972).

[116] Pitts, J. N., Jr., Letsinger, R. L., Taylor, R. P., Patterson, J. M., Recktenwald, G., Martin, R. B.: J. Am. Chem. Soc. *81*, 1068 (1959).
[117] Huyser, E. S., Neckers, D. C.: J. Am. Chem. Soc. *85*, 3641 (1963).
[118] a) Monroe, B. M., Weiner, S. A.: J. Am. Chem. Soc. *91*, 450 (1969);
b) Davidson, R. S., Younis, F. A., Wilson, R.: Chem. Commun. *1969*, 826.
[119] Loutfy, R. O., deMayo, P.: Chem. Commun. *1970*, 1040.
b) Zimmerman, H. E., Hull, V. J.: J. Am. Chem. Soc. *92*, 6515 (1970).
[120] a) Padwa, A., Bergmark, W., Pashayan, D.: J. Am. Chem. Soc. *91*, 2653 (1969);
b) Fischer, M.: Chem. Ber. *100*, 3599 (1967).
[121] a) Nelsen, S. F., Bartlett, P. D.: J. Am. Chem. Soc. *88*, 137 (1966);
b) Lankamp, H., Nanta, W. T., MacLean, C.: Tetrahedron Letters *1968*, 249.
[122] Schenck, G. O., Cziesla, M., Eppinger, K., Matthias, G., Pape, M.: Tetrahedron Letters *1967*, 193.
[123] Wagner, P. J.: Mol. Photochem. *1*, 71 (1969).
[124] Chilton, J., Giering, L., Steel, C.: J. Am. Chem. Soc., in press.
[125] Filipescu, N., Minn, F. L.: J. Am. Chem. Soc. *90*, 1544 (1968).
[126] Arnold, D. R.: Advan. Photochem. *6*, 301 (1968).
[127] Carless, H. A. J.: J. Chem. Soc., Perkin II *1974*, 834.
[128] Saltiel, J., Neuberger, K. R., Wrighton, M.: J. Am. Chem. Soc. *91*, 3658 (1969).
[129] Yang, N. C., Cohen, J. I., Shani, A.: J. Am. Chem. Soc. *90*, 3264 (1968).
[130] Kochevar, I. E., Wagner, P. J.: J. Am. Chem. Soc. *94*, 3859 (1972).
[131] Caldwell, R. A., Sovocool, G. W., Gajewski, R. P.: J. Am. Chem. Soc. *95*, 2549 (1973).
[132] Wagner, P. J., Lewitt, R. A.: J. Am. Chem. Soc. *95*, 3669 (1973).
[133] Turro, N. J., Dalton, J. C., Farrington, G., Niemczyk, M., Pond, D. M.: J. Am. Chem. Soc. *92*, 6978 (1970).
[134] Turro, N. J., Wriede, P. A., Dalton, J. C., Arnold, D., Glick, A.: J. Am. Chem. Soc. *89*, 3950 (1967).
[135] Turro, N. J., Wriede, P. A.: J. Am. Chem. Soc. *92*, 320 (1970).
[136] Yang, N. C., Hui, M. H., Bellard, S. A.: J. Am. Chem. Soc. *93*, 4056 (1971).
[137] Singer, L. A., Davis, G. A., Muralidharan, V. P.: J. Am. Chem. Soc. *91*, 897 (1969).
[138] Yang, N. C., Loeschen, R., Mitchell, D.: J. Am. Chem. Soc. *89*, 5465 (1967).
[139] Yang, N. C., Loeschen, R. L.: Tetrahedron Letters *1968*, 2571.
[140] de Mayo, P.: Accounts Chem. Res. *4*, 41 (1971).
[141] Chapman, O. L., Koch, T. H., Klein, F., Nelson, P. J., Brown, E. L.: J. Am. Chem. Soc. *90*, 1657 (1968).
[142] Eaton, P. E.: Accounts Chem. Res. *1*, 50 (1968).
[143] Wagner, P. J., Bucheck, D. J.: J. Am. Chem. Soc. *91*, 5090 (1969).
[144] Cookson, R. C., Hudec, J., Knight, S. A., Whitear, B.: Tetrahedron Letters *1962*, 79.
[145] Corey, E. J., Bass, J. D., Le Mahieu, R., Mitra, R. B.: J. Am. Chem. Soc. *86*, 5570 (1964).
[146] Dauben, W. G., Shaffer, G. W., Vietmeyer, N. D.: J. Org. Chem. *33*, 4060 (1968).
[147] a) Dilling, W. L., Tabor, T. E., Boer, F. P., North, P. P.: J. Am. Chem. Soc. *92*, 1399 (1970);
b) Loutfy, R. O., de Mayo, P.: Can. J. Chem. *50*, 3465 (1972).
[148] McCullough, J. J., Kelly, J. M., Rosmussen, P. W. W.: J. Org. Chem. *34*, 2933 (1969).
[149] Ruhlen, J. L., Leermakers, P. A.: J. Am. Chem. Soc. *89*, 4944 (1967).
[150] Eaton, P. E., Cole, T. W.: J. Am. Chem. Soc. *86*, 3157 (1964).
[151] Cargill, R. L., Crawford, J. W.: J. Org. Chem. *35*, 356 (1970).
[152] Cohen, S. G., Parola, A., Parsons, G. H.: Chem. Rev. *73*, 141 (1973).
[153] Cohen, S. G., Cohen, J. I.: J. Am. Chem. Soc. *89*, 164 (1967).
[154] Davidson, R. F., Lambeth, P. F.: Chem. Commun. *1968*, 511.
[155] Wagner, P. J., Kemppainen, A.: J. Am. Chem. Soc. *91*, 3085 (1969).
[156] Cohen, S. G., Litt, A. D.: Tetrahedron Letters *1970*, 837.
[157] Arimitsu, S., Masuhara, H.: Chem. Phys. Letters *22*, 543 (1973). — Arimitsu, S., Masuhara, H., Mataga, N., Subomura, H.: J. Phys. Chem. *79*, 1255 (1975).
[158] Saudner, M. R., Osborn, C. L., and Trecker, D. J.: J. Polymer. Sci., Part 1-A, Polymer Chem. *10*, 3173 (1972).

159) a) Atkins, P. W., Evans, G. T.: Mol. Phys. *27*, 1633 (1974);
b) Pedersen, J. B., Freed, J. H.: J. Chem. Phys. *62*, 1706 (1975);
c) Adrian, F. J.: J. Chem. Phys. *61*, 4875 (1974);
d) Adeleke, B. B., Choo, K. Y., Wan, J. K. S.: J. Chem. Phys. *62*, 3822 (1975);
e) Dobbs, A. J., McLauchlan, K. A.: Chem. Phys. Letters *30*, 257 (1975);
f) Wong, S. K., Wan, J. K. S.: J. Am. Chem. Soc. *94*, 7197 (1972).

160) For leading references, see Souto, M. A., Wagner, P. J., El-Sayed, M. A.: Chem. Phys. *6*, 193 (1974).

161) a) Fessenden, R. W., Schuler, R. H.: J. Chem. Phys. *39*, 2147 (1963);
b) Wan, J. K. S., Wong, S.-K., Hutchinson, D. A.: Accounts Chem. Res. *7*, 58 (1974).

162) a) Vyas, H. M., Wong, S. K., Adeleke, B. B., Wan, J. K. S.: J. Am. Chem. Soc. *97*, 1385 (1975);
b) Bargon, J., Siefert, K. G.: Ber. Bunsenges. Physik. Chem. *78*, 1180 (1974);
c) Choo, K. Y., Wan, J. K. S.: J. Am. Chem. Soc., *97*, 7127 (1975).

163) Roth, H. D., Lamola, A. A.: J. Am. Chem. Soc. *96*, 6270 (1974).

164) a) Cohen, S. G., Gutenplan, J. B.: Tetrahedron Letters *1968*, 5353; *1969*, 2125;
b) Davis, G. A., Carapellucci, P. A., Szoc, K., Gresser, J. D.: J. Am. Chem. Soc. *91*, 2264 (1969);
c) Bartholomew, R. F., Davidson, R. S., Howell, M. J.: J. Chem. Soc. C *1971*, 2804.

165) Cohen, S. G., Davis, G. A., Clark, W. D. K.: J. Am. Chem. Soc. *94*, 869 (1972).

166) Davidson, R. S., Santhanam, M.: J. Chem. Soc., Perkin II *1972*, 2355.

167) Chuang, T. J., Cox, R. J., Eisenthal, K. B.: J. Am. Chem. Soc. *96*, 6828 (1974).

168) Wagner, P. J., Jellinek, T., Kemppainen, A. E.: J. Am. Chem. Soc. *94*, 7512 (1972).

169) Scheve, B. J.: Ph. D. Thesis, Michigan State University, 1974.

170) Ersfeld, D. A.: Ph. D. Thesis, Michigan State University, 1975.

171) Rehm, D., Weller, A.: Ber. Bunsenges. Physik. Chem. *73*, 834 (1969).

172) Gutenplan, J. B., Cohen, S. G.: J. Am. Chem. Soc. *94*, 4040 (1972).

173) Evans, T. R.: J. Am. Chem. Soc. *93*, 2081 (1971).

174) Taylor, G. N., Hammond, G. S.: J. Am. Chem. Soc. *94*, 3684 (1972). — Stephenson, L. M., Hammond, G. S.: Pure Appl. Chem. *16*, 125 (1968).

175) a) Gutenplan, J. B., Cohen, S. G.: Tetrahedron Letters, *1972*, 2163;
b) Loutfy, R. O., Loutfy, R. O.: Can. J. Chem. *50*, 4052 (1972).

176) Carrol, F. A., McCall, M. T., Hammond, G. S.: J. Am. Chem. Soc. *95*, 315 (1973).

177) a) Clark, W. D. K., Litt, A. D., Steel, C.: J. Am. Chem. Soc. *91*, 5413 (1969);
b) Saltiel, J., Curtis, H. C., Metts, L., Miley, J. W., Winterle, J., Wrighton, M.: J. Am. Chem. Soc. *92*, 410 (1970);
Parker, C. A.: Chem. Commun. *1968*, 749.

178) a) Schuster, D. I., Weil, T. M., Halpern, M.: J. Am. Chem. Soc. *94*, 8248 (1972);
b) Loutfy, R. O., Yip, R. W.: Can. J. Chem. *51*, 1881 (1973).

179) a) Dedinas, J.: J. Phys. Chem. *75*, 181 (1971);
b) Saltiel, J., Curtis, H. C., Jones, B.: Mol. Photochem. *2*, 331 (1970);
c) Schuster, D. I., Brizzolara, D. F.: J. Am. Chem. Soc. *92*, 4357 (1970).

180) Brimage, D. R. G., Davidson, R. S., Lambeth, P. F.: J. Chem. Soc. (C) *1971*, 1241.

181) Barltrop, J. A., Carless, H. A. J.: J. Am. Chem. Soc. *93*, 4794 (1971).

182) Lam, H. M.-H.: M. S. Thesis, Michigan State University, 1975.

183) Bryce-Smith, D., Cox, G. B., Gilbert, A.: Chem. Commun. *1971*, 914.

184) Davies, A. G., Scaiano, J. C.: J. Chem. Soc., Perkin II *1972*, 2234.

185) Davidson, R. S., Steiner, P. R.: J. Chem. Soc., Perkin II *1972*, 1357.

186) a) Koch, T. H., Jones, A. H.: J. Am. Chem. Soc. *92*, 7503 (1970);
b) Schuster, D. I., Goldstein, M. D.: J. Am. Chem. Soc. *95*, 986 (1973).

187) Cohen, S. G., Parsons, G. H., Jr.: J. Am. Chem. Soc. *92*, 7603 (1970).

188) a) Schuster, D. I., Weil, T. M.: J. Am. Chem. Soc. *95*, 4091 (1973);
b) Singer, L. A., Brown, R. E., Davis, G. A.: J. Am. Chem. Soc. *95*, 8638 (1973).

189) Schaffner, K., Jeger, O.: Tetrahedron *30*, 1891 (1974).

190) Jaffé, H. H., Orchin, M.: Theory and applications of ultraviolet spectroscopy, p. 181. New York: J. Wiley and Sons, Inc. 1962.

[191] Leonard, N. J., Owens, F. H.: J. Am. Chem. Soc. *80*, 6039 (1958).
[192] Hlavka, J. J., Bitha, P.: Tetrahedron Letters *1966*, 3843.
[193] Iwasaki, S., Schaffner, K.: Helv. Chim. Acta *51*, 557 (1968).
[194] Sheehan, J. C., Wilson, R. M., Oxford, A. W.: J. Am. Chem. Soc. *93*, 7222 (1971).
[195] Cookson, R. C., Hudec, J.: J. Chem. Soc. *1962*, 429.
[196] Zimmerman, H. E., Sandel, V. R.: J. Am. Chem. Soc. *85*, 915 (1963).
[197] Barltrop, J. A., Thomson, A.: J. Chem. Soc. (C) *1968*, 155.
[198] McGrath, J. M.: Ph. D. Thesis, Michigan State University, 1972.

Received July 17, 1975

Triplet States from Azo Compounds

Professor Dr. Heinz Dürr and Dr. Bernd Ruge

Fachbereich 14 der Universität des Saarlandes, Organische Chemie, D-6600 Saarbrücken

Contents

1. Introduction

The photolysis of azo compounds with ultraviolet light can, in principle, lead to
1) unimolecular and
2) bimolecular reactions.

The unimolecular reactions are the predominant ones, consisting of: *cis-trans*-isomerisation, tautomerisation and elimination of nitrogen. While these reactions mainly occur in the azo alkane series, with aromatic azo compounds, cyclisation and photoreduction are observed in addition to these reactions.

In this article reactions of both acyclic and cyclic azo compounds with regard to triplet intermediates are collected. Typical heterocyclic systems as well as triazines *etc.* are excluded.

2. Spectroscopy

2.1. Absorption Spectra of Azo Compounds

The azo group –N=N– is very important as an essential entity of organic dyes. On the other hand interesting photochemistry of azo compounds is to be expected since these molecules contain a chromophore absorbing the uv light as well as a leaving group, *e.g.* nitrogen, of low energy. In addition to the bands arising from π-electrons of the N=N-double bond, the bands due to lone pair electrons on nitrogen further complicate the spectra of azo compounds.

In Tables 1, 2 and 3 ultraviolet absorption spectra of acyclic or cyclic, aliphatic and aromatic azo compounds are compiled.

Table 1. $n \rightarrow \pi^*$-Bands of aliphatic azo compounds

Compound	$\tilde{\nu}_{max}$ [cm^{-1}]	λ_{max} [nm]	ε_{max}	Ref.
C_{2h}-Symmetry:				
trans-$CH_3N{=}NCH_3$	28000	357	13	6)
trans-$(CH_3)_2CHN{=}NCH(CH_3)_2$	28000	357	17	6)
trans-$(CH_3)_3CN{=}NCH(OCOCH_3)(CH(CH_3)_2)$	28000	357	19	6)
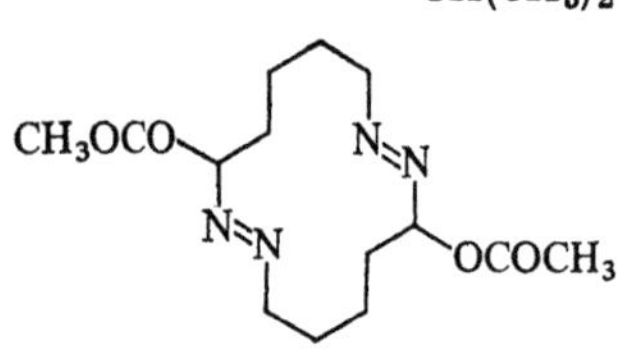	28000	357	60	6)
Azobiscyclohexyl cyanide	28600	350	11	7)
2,2'-Azobis-2-methylpropionitrile	29000	345	12	7)
EtPhCH–N=N–CHEtPh	27800	359	–	7)

Table 1 (continued)

Compound		$\tilde{\nu}_{max}$ [cm^{-1}]	λ_{max} [nm]	ε_{max}	Ref.
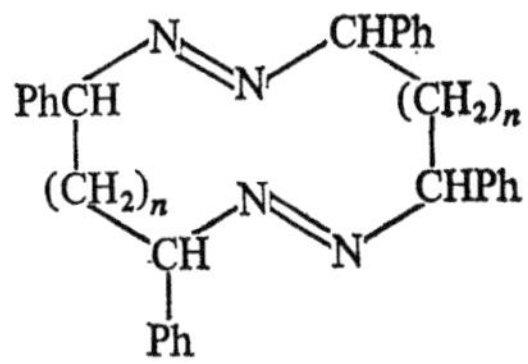		27800	360	—	[7]
$CH_2{=}C(CH_3){-}N{=}N{-}CH_3$		25900	386	6060	[7]
$CH_3CO{-}N{=}N{-}COCH_3$		22100	452	30	[7]
C_{2v}-Symmetry:					
Cis-$CH_3N{=}NCH_3$		28400	352	240	[6]
Diazirine		32400*	309	176	[7]
3-Methyldiazirine		29300	341	470	[7]
		31000*	322	—	[7]
3,3-Dimethyldiazirine		27700	360	180	[7]
		29000*	345	—	[7]
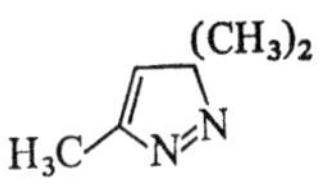	*1*	38900	257	2250	[7]
		28600	350	270	[7]
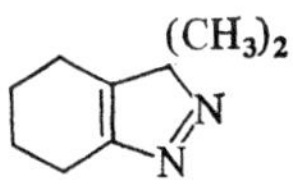	*2*	36800	272	3530	[7]
		28800	347	300	[7]
	*3*	29500	349	400	[6]
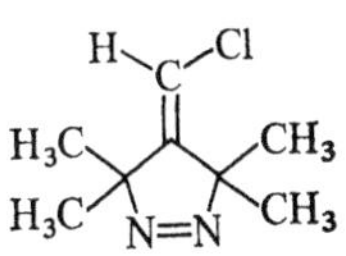	*4a*	31000	333	280	[6]
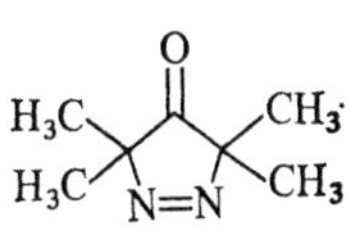	*5*	28000	357	250	[6]
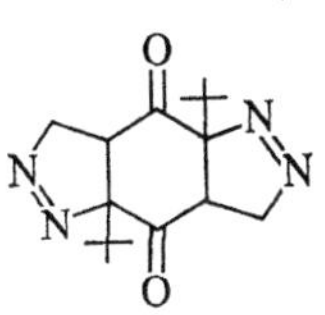	*6*	30500	328	—	[6]

Table 1 (continued)

Compound		$\tilde{\nu}_{max}$ [cm^{-1}]	λ_{max} [nm]	ε_{max}	Ref.
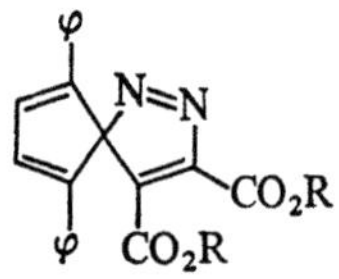	7	27900	365	—	8,9)
	8a	27000	371	150	6)
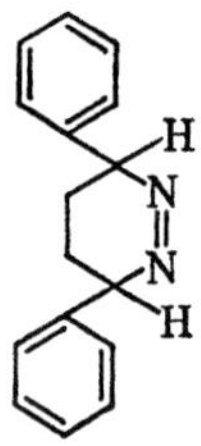	9	25900	408	770	6)

Table 2. Spectra of substituted *trans*-azo benzenes p—R—$C_6H_4N{=}NC_6H_5$ [6)]

R =	$n\to\pi^*$-Band			Long wavelength $\pi\to\pi^*$-band		
	$\tilde{\nu}_{max}$ [cm^{-1}]	λ_{max} [nm]	lg ε_{max}	$\tilde{\nu}_{max}$ [cm^{-1}]	λ_{max} [nm]	lg ε_{max}
H	22300	448	2.61	31600	327	4.35
$N(CH_3)_3$ J	22500	444	2.71	31600	327	4.3
CH_3	22200	450	2.83	30000	334	4.37
OCH_3	22500	444	2.8	29000	345	4.31
NO_2	22000	455	2.83	30000	334	4.35
$N(CH_3)_2$	21500[1)]	465	—	24500	408	4.36
$N{=}NC_6H_5$	22000	455	3.4	27900	359	4.6

[1)] Calculated.

The aliphatic azo compounds show a characteristic $S_0\to S_1$ absorption in the region of 320—380 nm, whereas aromatic azo compounds absorb between 430 and 460 nm. These bands are due to $n\to\pi^*$-transitions of the azo compounds concerned. In acyclic azo compounds the $n\to\pi^*$-band is dependent on substitution, whereas in aromatic azo compounds no such influence is noticeable. However $\pi\to\pi^*$-bands in aromatic azo compounds respond quite strongly to substitution while, the intensity of $n\to\pi^*$-bands is only weakly affected by substitution. The $n\to\pi^*$-bands

Table 3. Spectra of *trans*-diaryl azo compounds [6)]

Compound	$n \to \pi^*$-Band			1. $\pi \to \pi^*$-Band				
	$\tilde{\nu}_{max}$ cm^{-1}	λ_{max} nm	$\lg \varepsilon_{max}$	$\tilde{\nu}_{0-0}$ cm^{-1}	λ_{max} nm	$\tilde{\nu}_{max}$ cm^{-1}	λ_{max} nm	$\lg \varepsilon_{max}$
Ph—N=N—Ph	22500	444	2.65	28500	351	31500	317	4.35
Ph—N=N—1-Napht	22100	453	2.95	26000	385	27000	371	4.08
Ph—N=N—2-Napht	22400	447	2.96	26000	385	27000	371	3.8
1—Napht—N=N—1-Napht	21600	463	3.17	23500	426	25000	400	4.22
1—Napht—N=N—2-Napht	22000	455	3.18	25000	400	26300	381	4.29
2—Napht—N=N—2-Napht	22500	444	3.18	25500	393	25500	392	4.15

Ph = Phenyl
Napht = Naphtyl

of both aliphatic and aromatic azo compounds show no structure, usually broad bands are obtained. Low temperature as well as solutions in polyethylene do not increase the fine structure of the spectra. In contrast to $n \to \pi^*$-bands the $\pi \to \pi^*$-bands in aromatic azo compounds show fine structure. If the azo group is however incorporated in rigid rings such as *3* or *8* the absorption sharpens and fine structure can be seen (see Fig. 1).

From these spectra the O—O-band, *e.g.* the $S_0 \to S_1$-band, can easily be derived, whereas in flexible or acyclic azo compounds the energy of the lowest singlet state must be taken from the onset of the broad $n \to \pi^*$-band. According to group theory [6)] the lowest energy transition being of the $n \to \pi^*$-type is symmetry forbidden, for *trans*-azo compounds ($n_s \to \pi^*$), which have C_{2h}-symmetry. For the *cis*-azo compounds (C_{2v}-symmetry) the $n \to \pi^*$-band is symmetry allowed ($n_a \to \pi^*$).

HMO-calculations and a MIM (Molecules in Molecule) [6,6a,10)] calculation are shown in Figs. 2, 3 and 4.

The $n \to \pi^*$- and the $\pi \to \pi^*$-bands are indicated in Fig. 2. The different symmetry of the highest MO can also be seen from Fig. 2. Other calculations have been carried out [10–14)], the most complete study is however an *ab initio* calculation of HN=NH [12)]. In a recent *ab initio* calculation [15)] a dependence of C—NN-angle on the energy of the $^1(n \to \pi^*)$-transition of cyclic azo compounds was found. Increasing the N—N—C-angle leads to a decrease in the $n \to \pi^*$-excitation energy, *e.g.* to a bathochromic shift of the band concerned. The high intensity of the forbidden $n \to \pi^*$-band in *tans*-azo compounds is due to a coupling with the $\pi \to \pi^*$-transition. Thus energy is transferred to the $n \to \pi^*$-transition, increasing the intensity of the latter.

2.2. Emission Spectroscopy

Emission spectroscopy is usually used to determine the energy of the lowest singlet state E_S (fluorescence) and the lowest triplet state E_T (phosphorescence). However no fluorescence could be observed for either acyclic or cyclic *trans*-azo

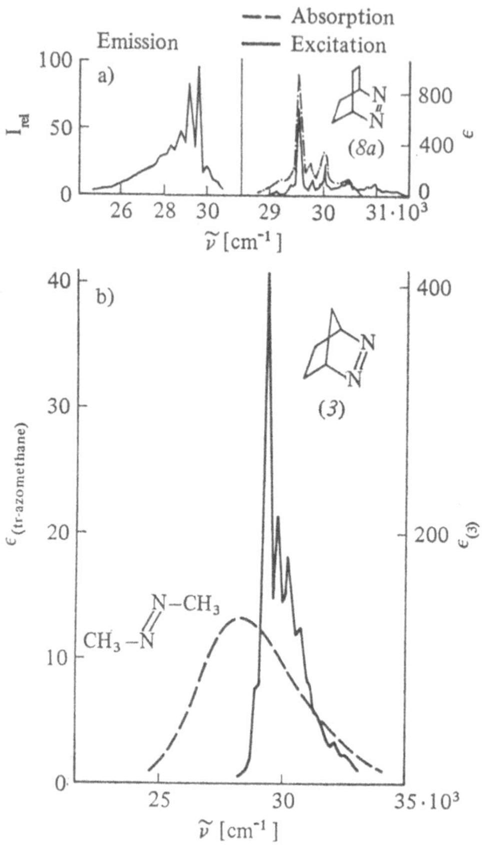

Fig. 1a. a) Gas spectra of (*8a*); b) Absorption of —— (*3*) in isooctane; ---- *trans*-azomethane-in hexane [6)]

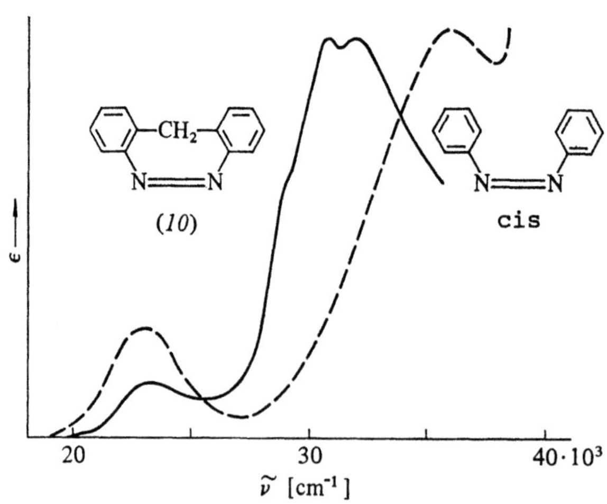

Fig. 1b. Absorptions of *cis*-azobenzene ----- and o.o'azodiphenylmethane (*10*) — in methylcyclohexane/isopentane at 77 K

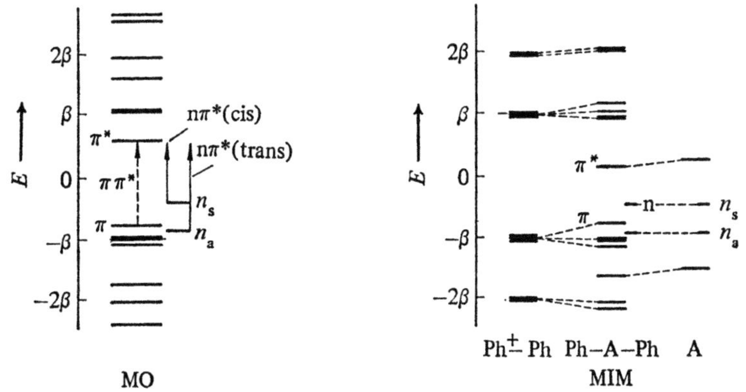

Fig. 2. Energy levels of azobenzene in Hückel-MO and MIM-approximation

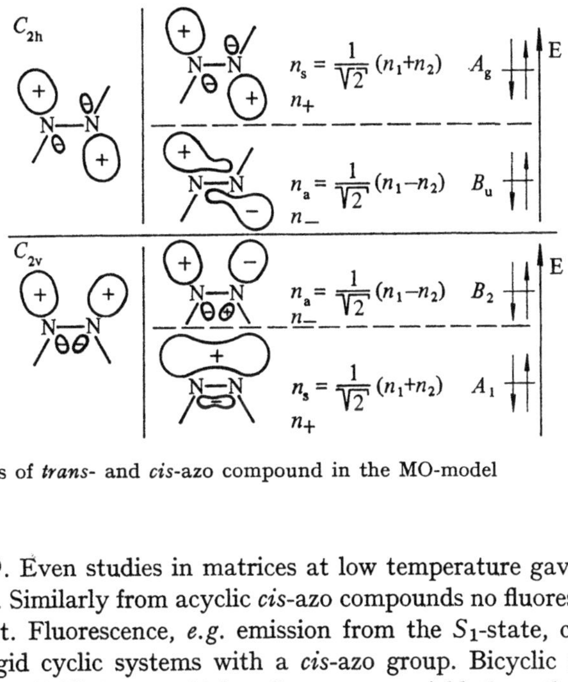

Fig. 3. *n*-Orbitals of *trans*- and *cis*-azo compound in the MO-model

compounds [5,6]. Even studies in matrices at low temperature gave no indication of fluorescence. Similarly from acyclic *cis*-azo compounds no fluorescence has been detected as yet. Fluorescence, *e.g.* emission from the S_1-state, can however be observed in rigid cyclic systems with a *cis*-azo group. Bicyclic azo-compounds such as *3* and *8* [16,17] have a higher fluorescence yield than the pyrazolines *4* and *6*. The o-diazines generally exhibit fluorescence [19,20].

2.3. Triplet Energies

For a photochemical reaction it is important to know the energy of the triplet state. The triplet energy of a certain compound is most easily determined by measuring the 0′→0-band of phosphorescence. For acyclic or cyclic azo compounds no phosphorescence has been observed so far. The only exception to this are phthalazine *11* [21] and the diaza[2.2]spirene *7* [22]. Preliminary phosphores-

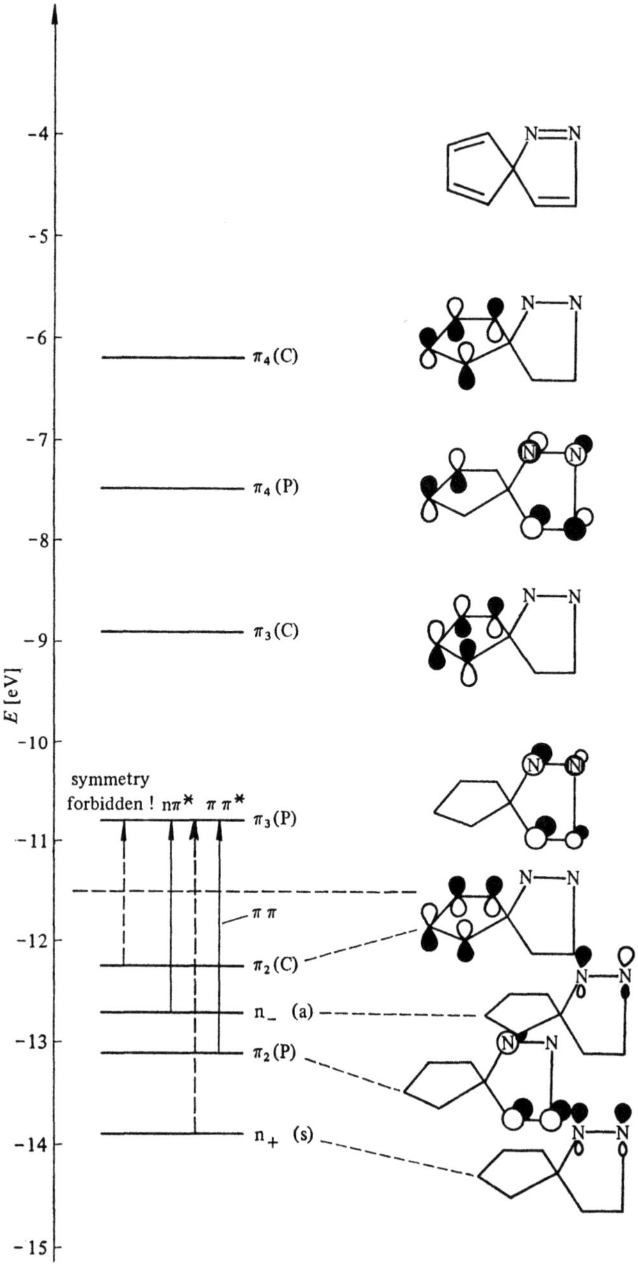

Fig. 4. Nature of excited state of diaza [2.2] spirene from an extended Hückel-calculation

cence studies on *7* also point to emission from a triplet state. From the emission maximum the energy of the triplet can be derived to be $E_T > 54$ kcal/mole [22].

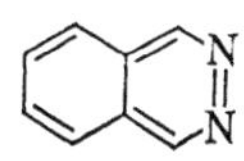

11

A *second method* for observing phosphorescence spectra is to populate the triplet *via* energy transfer with sensitizers. With this sensitized phosphorescence a very inefficient intersystem crossing step can be overcome. The bicyclic azo compound *8a* and its 3,6-dimethylanalog *8b* appeared ideal for such a study since they show little photochemistry from the triplet state. However this investigation was not successful [23] (see also [24]). The triplet energy of the $^3(n\pi^*)$ and $^3(\pi\pi^*)$ state of pyridazine [25a] and benzo[c]cinnoline *12* [25b] could be determined from the crystal absorption spectra at 4.2 K.

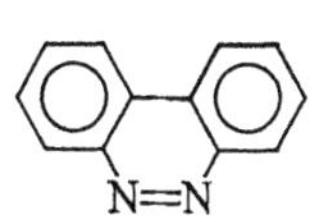

12

Another possibility to determine the triplet energy of a molecule is to monitor the efficiency of energy transfer. This is possible if the molecule undergoes a definite reaction. This is the case with azo compounds which can loose N_2 very efficiently. Diazabicycloheptene *3* is an ideal case, because here nitrogen elimination occurs with unit efficiency [23]. The triplet sensitized quantum yield of nitrogen evolution is equal to the product of the probability of intersystem crossing multiplied by the probability of triplet energy transfer to *3*. If the triplet of the donor is about 3 kcal/mole higher than that of the acceptor the energy transfer rate is at the diffusion controlled limit. If the energy of the donor decreases, the energy transfer rate falls drastically or in other words the quantum yield of reaction is also reduced. In the photolysis of *3*, on going from the sensitizer Michler's ketone ($E_T = 61.0$ kcal/mole) to 2-acetonaphthone ($E_T = 59.3$ kcal/mole) the quantum yield drops from unity to almost zero. This places the triplet state of *3* at about 60 kcal/mole. If the triplet acceptor does not undergo a well defined reaction, the energy of the triplet state can be calculated if the rate constant of triplet energy transfer k_t can be measured. 1,1-Azo-isobutane quenches the biacetyl triplet placing the triplet energy of 1,1-azoisobutane at about 55 kcal/mole. More recent work gave a similiar value of the order of 54 ± 1 kcal/mole [27].

From similiar studies of the *cis-trans*-isomerisation of azobenzene a maximum value of about 45 kcal/mole was derived [28]. However no clear distinction between a $^3n\pi^*$- or a $^3\pi\pi^*$ was made.

Care should however be taken in using sensitization or quenching experiments to derive the triplet energy of azo compounds. Steric interaction of bulky *trans*-azo compounds like azo-t-butane can greatly reduce the energy transfer rates thus giving rise to wrong values for triplet energies of azo compounds [28]. No such problems seem to arise in *cis*-azo compounds [29]. A series of triplet energies for azo compounds is collected in Table 4.

For a discussion of the triplet state of pyridazines see [30].

Table 4. Excited state energies of selected azo compounds [5]

Compound	E_S kcal	E_T kcal	E_{S-T} [1] kcal
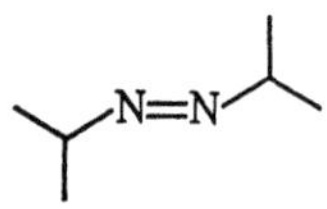	<68	54	<14
3 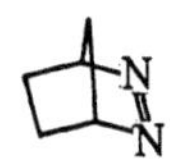	84.3	60 ± 1	24
8a 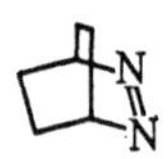	76.0	54.5 ± 1	21.5
13 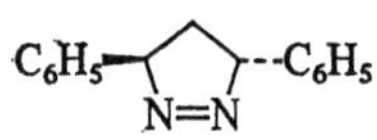	80 [2]	55	25
	90.5 [3]	69.5 [3]	21.0
	77.6 [3]	47.8 [3]	29.8
	–	<45 [28]	–
7 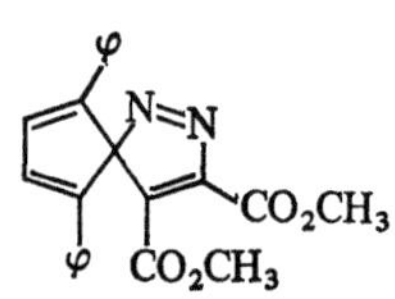	–	>54 [22]	–

[1] Energy gap from lowest excited singlet to lowest triplet state.
[2] Estimated from the 358 nm absorption band origin in 3,6-bis(p-methoxyphenyl)-1-pyrazoline.
[3] Calculated values from Ref. [12].

2.4. Photoelectron Spectroscopy

Photoelectron spectroscopy is a very valuable tool in measuring the different energy levels of a molecule. This is however based on the assumption that the ionization potentials which are determined experimentally are directly correlated with the MO's of a certain molecule, *i.e.* Koopman's theorem must be valid. The few azo compounds studied so far already allow a clear assignment of the orbitals concerned and a deeper insight into the ordering of the n and π-states in azo compounds.

The photoelectron spectra and the assignment of the various bands for *trans*- and *cis*-azo compounds are collected in Table 5. The assignments are usually made with the aid of semiempirical calculations such as MINDO or CNDO procedures.

Table 5. Photoelectron spectra of *trans*- and *cis*-azo compounds (all bands are given in eV)

Compound	n	π	n	σ	Ref.
CH_3–N=N–CH_3	n_s (n_+)		n_a (n_-)		
	8.98	11.38	12.3	13.5	[31]
		11.84		14.6	
		11.53		15.8	
		11.68			
		12.0			
 *3*	n_a (n_-) 8.96	11.53	n_s (n_+) 11.91		[32]
 *8a*	8.32	10.70	complex bands		[32]
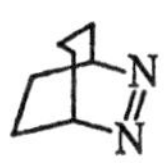 *14*	8.07	10.60	11.2–11.6[1])		[32]
 *15*	8.02	(10.52)[1])	complex bands		[32]

[1]) Complex bands.

From the photoelectron spectrum of *trans*-azomethane the ordering of the different states is

$$n_+ \equiv (n_s) < \pi < n_- ,$$

e.g. the transition with lowest energy in trans-azomethane is expected to be the symmetric $n_s \rightarrow \pi^*$ transition [31]. The different states of azomethane, together with the corresponding orbitals obtained by MINDO/1 [31] are shown in Fig. 5.

More refined calculations give the same assignments [31]. The photoelectron spectrum of diazirine has also been studied [33]. In diazirine as well as in the bicyclic

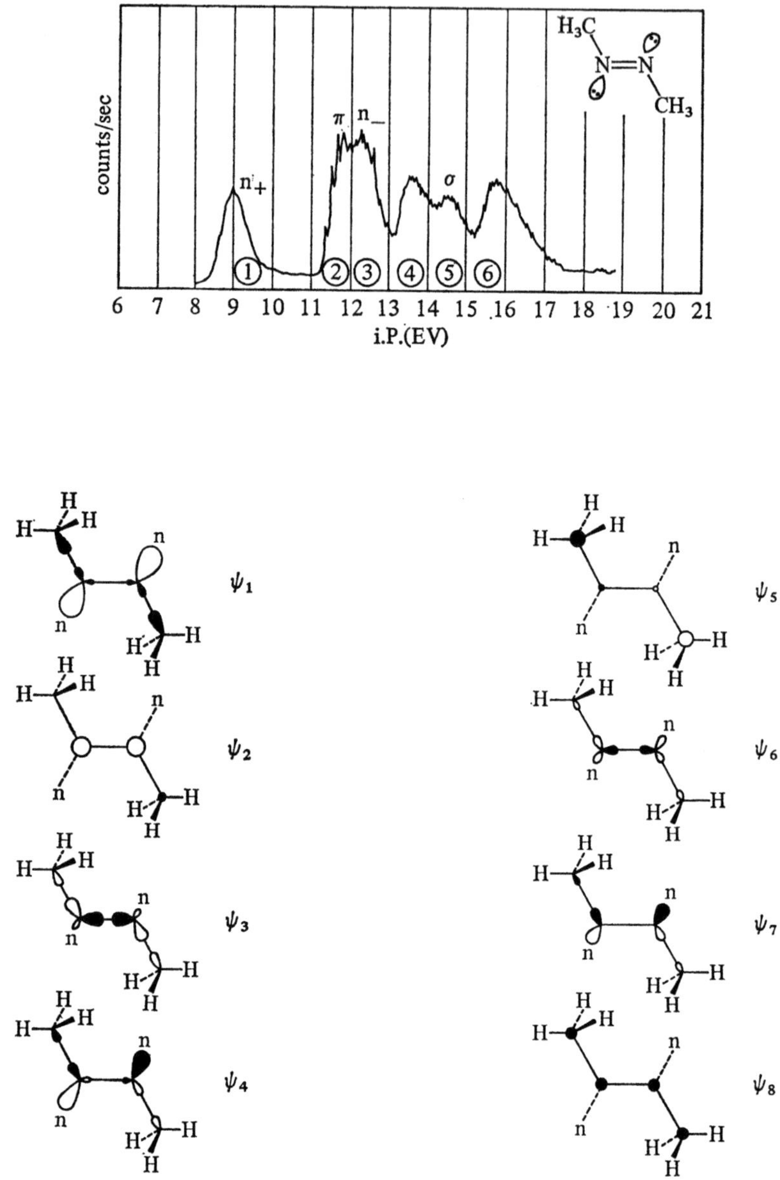

Fig. 5. Photoelectron spectrum and molecular orbitals of *trans*-azomethane [31]. *Abszisse:* ionisation potential in eV; *Ordinate:* counting rates in arbitrary units

azo alkanes (Table 5) the two n-states are reversed, because of C_{2v}-symmetry. The lowest energy band in the PE-spectrum is now the anti-symmetric n_a-band. As Table 5 demonstrates, this n_a-band is in the same region as the n_s-band for trans-azomethane. With decreasing strain in the bicyclic azo compound the n_a-band is displaced to lower values [32].

2.5. ESR-Spectroscopy

ESR spectroscopy can generally be used to study the multiplicity of the radicals formed by photolysis of azo compounds. Very few studies of this type have however been reported. Photolysis of the azo compounds *16, 17* and *18* in isopentane/methylcyclohexane in a matrix did not give rise to the radical pair but instead hydrazyl radicals *19* were detected [34]. These were formed by attack of the initially produced radicals on solvent, which then adds as a radical to the azo compound to give *19*.

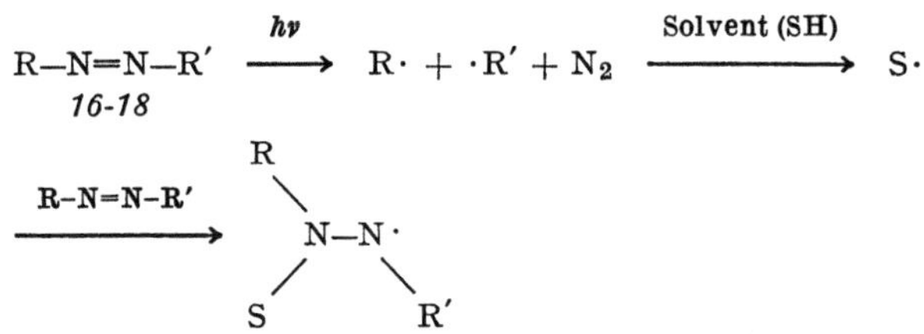

19

	R	R'
16	φ C(CH₃)₂	φ
17	φ	$CO_2C_2H_5$
18	φ	$C(CH_3)_2CH{=}CH_2$

Bartlett studied the photolysis of the azo compound *20* with ESR spectroscopy [35]. A triplet ESR signal could be detected at low temperature, which is thought to be due to a triplet radical pair *20'*. This radical pair is thought to be held apart by a nitrogen molecule at about 7 Å. Warming up the system produces the ESR signal of the monoradical. This result is however in contradiction with recent CIDNP studies (see below).

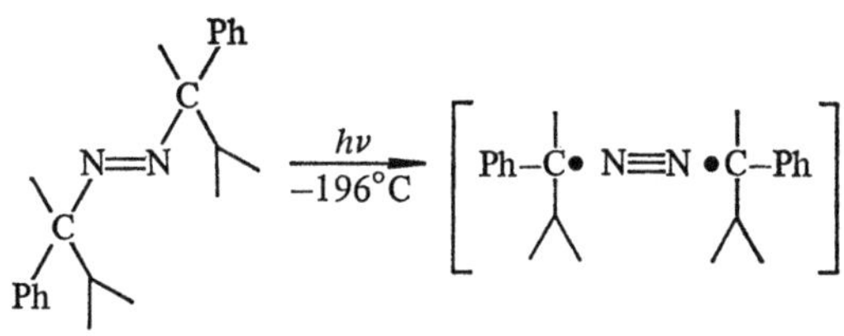

20 *20'*

An ESR study of the photolysis of the 3*H*-indazole *21* revealed at low temperature the formation of a triplet biradical *22*. This can be trapped with 1,1-diphenylethylene to give — after warming to 140 K — the 1,5-diradical *23* [36].

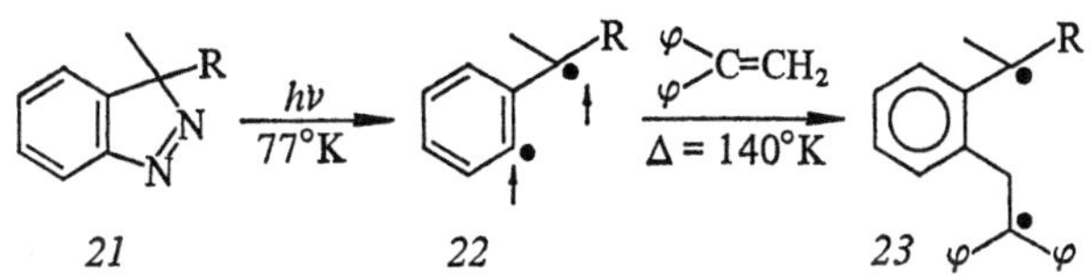

21 *22* *23*

R = CN, CO_2CH_3, OCH_3

These results were however questioned recently by Schrader [37]. A singlet *22* was thought to be formed first eliminating nitrogen to give a singlet biradical *22* (↑↓) which then undergoes intersystem crossing to triplet *22* (↑↑).

2.6. CIDNP-Studies

To study the photolysis of azo compounds, CIDNP was only recently introduced in the field of photochemistry. The CIDNP-effect consists of generating a geminate radical pair which still remembers the spin state of its precursor. So the multiplicity of the precursor can be determined from enhanced absorption or emission signals in azoalkane photolysis. The benzophenone sensitized photolysis of diazirine in deuteriochloroform leads to the triplet azo compound *24* which decomposes under elimination of a ground state nitrogen molecule and a triplet methylene [38]. This abstracts deuterium from deuteriochloroform to form the geminate radical pair *25*. This can now recombine to give *26* or dissociate to afford the free radical products.

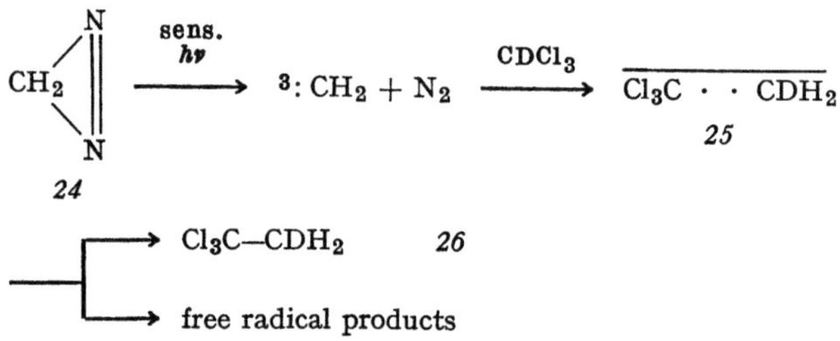

26 shows an enhanced absorption for the methylene protons (the direct photolysis produces an enhanced emission signal) clearly indicating the involvement of a triplet in this reaction (see Fig. 6).

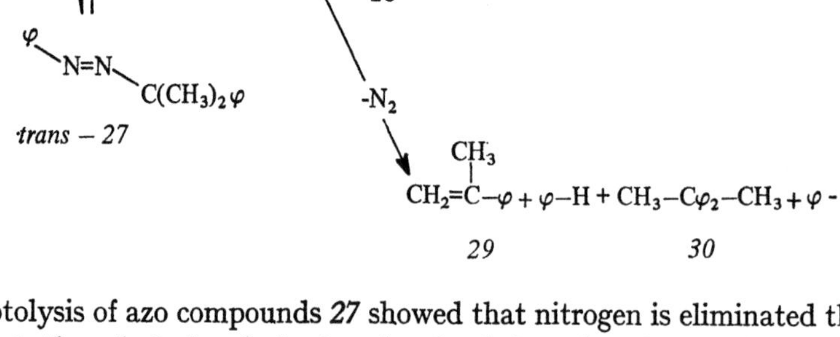

Photolysis of azo compounds *27* showed that nitrogen is eliminated thermally, but that the photochemical step involved is only *cis-trans*-isomerisation of *27* [39,52].

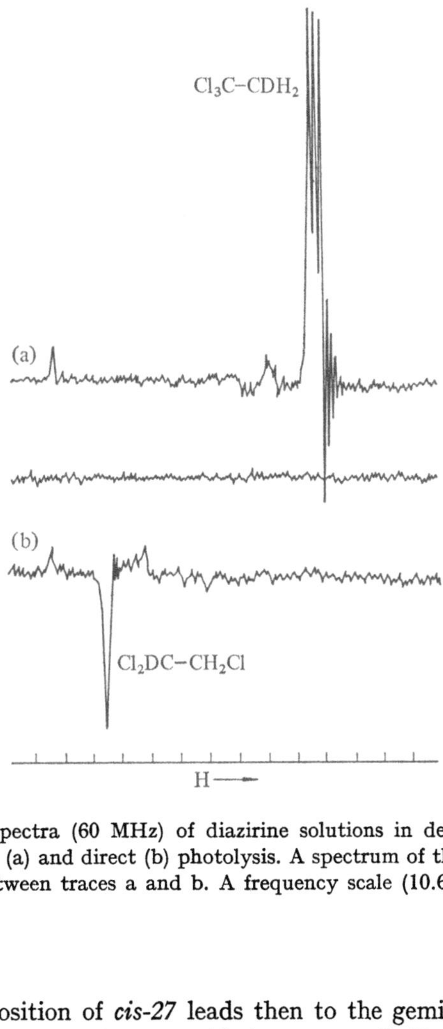

Fig. 6. Proton NMR Spectra (60 MHz) of diazirine solutions in deuteriotrichloromethane during triplet-sensitized (a) and direct (b) photolysis. A spectrum of the same region without irradiation is shown between traces a and b. A frequency scale (10.6 Hz between markers) is given below trace b

Thermal decomposition of *cis-27* leads then to the geminate radical pair *28* giving the products α-methylstyrene *29*, benzene and *30*. The main products after isolation are biphenyl (20%) and dicumyl (48%). This study shows that the decomposition of azo compounds proceeds *via* a diazenyl radical in the singlet reaction (see also [40)]). No such studies are available at the present time for the triplet.

3. Reactions of Triplet Azo Compounds

3.1. *Cis-trans*-Isomerisation of Azo Compounds

Acyclic azo compounds are obtained by usual routes only as *trans*-isomers. Photochemical isomerisation provides an easy and effective way to the *cis*-isomers. The *cis*-azo compounds are however thermally quite labile and decompose with elimination of nitrogen [41)], some of them even at —120 °C [5)]. Activation energies for the thermal decomposition of *cis*-azo-alkanes are of the order of 20 kcal/mole. Theoretical calculations for this process are available [42]. *Cis*-azo benzene cannot be planar any more, because of steric hindrance resulting from the *ortho*-hydrogens. The azo group is planar but the benzene rings are tilted away from each other by an angle of 56° [6]. The mechanism of the photochemical *cis-trans*-isomerisation of azo compounds is not completely clear. The photochemical isomerisation even at low temperature is very fast, *i.e.* no large barriers can be present [5]. *Cis*-azomethane can be prepared by *cis-trans*-isomerisation at —196 °C. In the *gas phase*, photolysis of azo compounds leads mainly to *nitrogen elimination*. On the other hand in the *condensed phase cis-trans-isomerisation* is the dominating process [43,44].

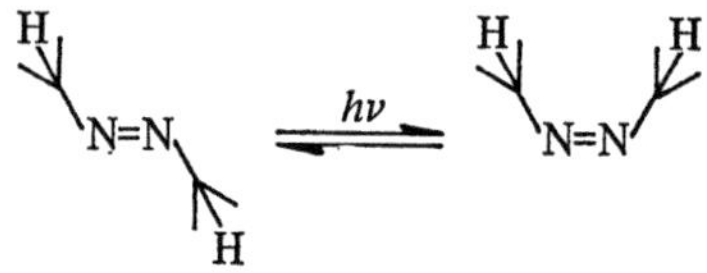

Quantum yield
of isomerisation

starting from: *trans*-azoisopropane: $\Phi^{trans}\,cis = 0.38$; $\Phi^{trans}\,trans = 0.51$
cis-azoisopropane: $\Phi^{cis}\,cis = 0.52$; $\Phi^{cis}\,trans = 0.56$

The quantum yields for *cis-trans*-isomerisation of azo-isopropane are almost equal starting either from *cis*- or *trans*-azoisopropane [44]. The *excited state* for isomerisation of azo alkanes must therefore have a very low barrier. Since the $n\pi^*$-state of azo alkanes still possesses considerable double bond character, isomerisation must occur from the $^3\pi\pi^*$-state [45]. Since triplet sensitizers do not effect nitrogen elimination [46] the most likely candidate for *cis-trans*-isomerisation is a $^3\pi\pi^*$-state. The high energy of the $^1\pi\pi^*$-state adds further weight to that argument. Thus one can conclude (1) that *cis-trans*-isomerisation of azo alkanes most probably proceeds from a $^3\pi\pi^*$-state, whereas (2) fragmentation most probably arises from a $^1n\pi^*$-state [5].

Aromatic azo compounds show a clear wavelength dependence of *cis-trans*-isomerisation. The results of different groups on the photostationary state of azo benzene are however quite contradictory [24,28,47,48]. *Cis*-azobenzenes revert easily to the more stable *trans*-isomers. This process has been shown to have an activation energy of about 20 kcal/mole. The quantum yields for *cis-trans*-isomerisation differ starting from *cis*- or *trans*-azo benzene [50].

There exists also a discrepancy in the data obtained for the sensitized photolysis of azo benzene. The different results of Jones and Hammond [28] for direct and sensitized photolysis most probably imply a $^3\pi\pi^*$-state in the sensitized and a singlet state in the direct photolysis. However the findings of Fischer [24,47] as well as theoretical calculations [11] and spectroscopic studies [51], again suggest a triplet state.

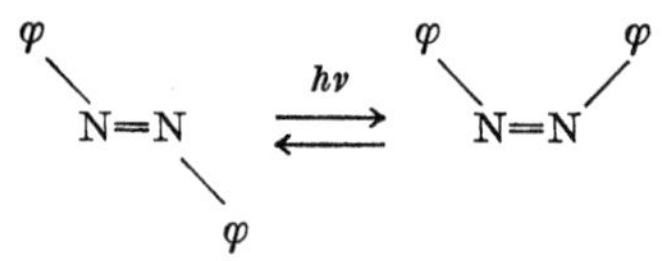

irradiation of potostationary state:	$\pi\pi^*$-band [28]	$n\pi^*$ [28]	sensitized	
trans/cis-ratio:	4/1	2/1	60/1 [28]	4/1 [24]

quantum yields Φ of *cis/trans*-isomerisation [50]

starting from	$n\pi^*$-excitation	$\pi\pi^*$-excitation
cis	0.70	0.44
trans	0.20	0.10

In a recent paper Porter and Marett [52] could demonstrate that photochemical *cis-trans*-isomerisation of optically active N-phenyl-N'-(2-phenyl-2-butyl)-diazene *31* as well as that of the 2-propyl-analog *32* occur at low temperature (—78°C) without nitrogen elimination. This *cis*- or *trans*-isomer produced predominantly depends on the wavelength used.

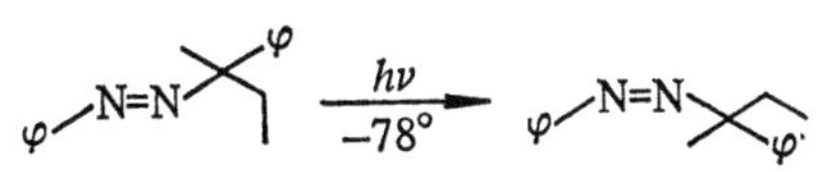

31 *33*

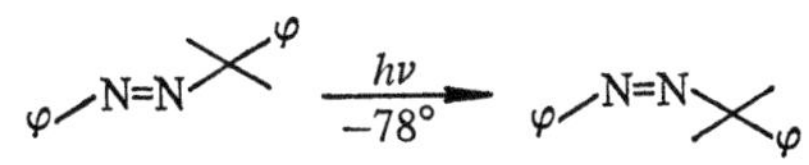

32 *34*

Optically active *31* during the *cis-trans*-isomerisation shows no racemization. *Cis-31* or *-32* already decompose thermally at room temperature. Therefore photolysis of *31* or *32* (room temperature) giving nitrogen evolution, must arise exclusively from thermal decomposition of the *cis*-azo compound. For *cis-trans*-isomerization of *31* and *32* two routes are suggested:

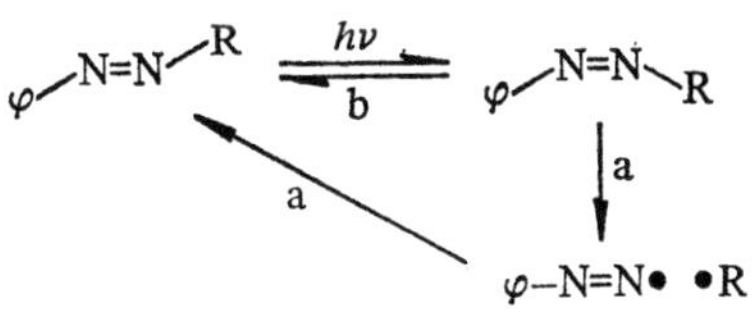

(1) a direct interconversion according to path b) or (2) one-bond scission and formation of a radical pair which can undergo internal return faster than racemization. The actual path followed depends obviously on the stability of the radical pair formed.

3.2. Nitrogen Elimination and Spin Correlation Effects

Early studies on the thermal decomposition of azo compounds seemed to indicate a concerted two-bond scission of the C—N-bonds (a) [53].

$$R\text{—}N{=}N\text{—}R \longrightarrow R\cdot \quad N_2 \quad \cdot R \quad \text{two-bond (a)}$$
$$R\text{—}N{=}N\text{—}R \longrightarrow R\text{—}N{=}N\cdot \quad \cdot R \quad \text{one-bond (b) scission}$$

However it was soon recognized that a one-bond scission (b) is in better agreement with experimental data [54]. Although a lot of work has been done in the field of photochemical decomposition of azo compounds the question of one-bond versus two-bond scission has only recently received more detailed attention. The photochemical elimination of nitrogen from an azo compound can in principle proceed in the following way:

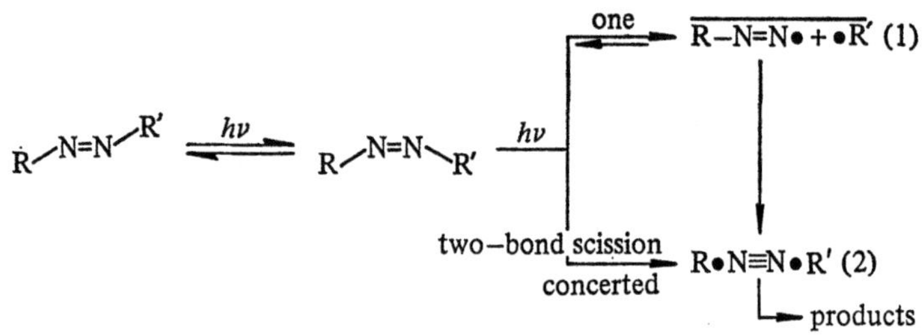

After excitation and population of the $n\pi^*$ or $\pi\pi^*$ state the excited molecule can undergo one-bond scission after *cis/trans*-isomerisation according to Eq. (1). Thus a radical R· and a diazenyl radical are formed. These radicals can either recombine to starting material (see above) or decompose into two radicals R·, R'· and a nitrogen molecule. The concerted fragmentation according to Eq. (2) is less probable on the basis of recent results. The elimination of nitrogen occurs as the experimental data demonstrate in general from a singlet excited state in acyclic azo compounds. The $^3\pi\pi^*$ state leads to *cis/trans* isomerisation only. In cyclic azo compound where no *cis/trans* isomerisation is possible, in small and medium rings elimination of nitrogen can arise from both the singlet and the triplet excited state.

Since the intersystem crossing step in azo compounds populating the triplet is not very efficient, the triplet state is populated in general by energy transfer.

Another important feature in azo compound photolysis is the *spin correlation effect*. The conservation of electron spin as well as energy considerations require that an excited *singlet* azo compound should decompose to ground state nitrogen and a *singlet biradical* or *singlet radical pair*. A *triplet* excited state however produces nitrogen and a *triplet biradical* or *radical pair*. Since bond formation of the triplet biradical or radical pair cannot occur before intersystem crossing to a singlet, a reaction faster than spin inversion can compete efficiently with this process. For singlet biradicals or radical pairs no such restrictions hold, so that no competing process should interfere.

Under these conditions photochemical decomposition from the singlet or the triplet state of azo compounds should lead to a different product ratio. This phenomenon is called the *spin correlation effect*. The simple scheme representing the photodecomposition of azo compounds (see above) has to be extended further if the complete mechanism has to be dealt with.

Spin conservation immediately after bond scission of the azo compound leads to an intimate or correlated radical pair *35*, which can diffuse from the *solvent cage* to give a free radical pair *36*. Two routes are also possible for the less probable concerted process *37*→*38*. Products formed from the intimate radical

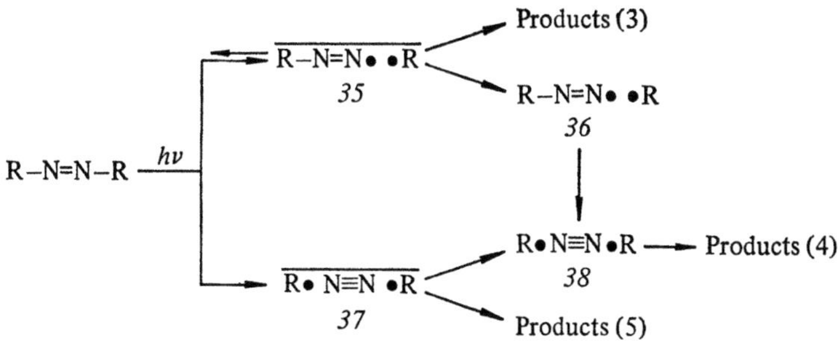

pairs *35* (or *37*) in the solvent cage still remember the original spin state from which they were generated. Only the cage products should therefore show a spin correlation effect.

In this chapter singlet reactions will always be compared with the triplet reactions to point out the spin correlation effect whenever possible.

In Table 6 quantum yields for direct and sensitized decomposition of azo compounds are collected.

Table 6. Nitrogen quantum yields for photolysis of acyclic azo compounds in nonpolar solvents

R in RN=NR[1]	Φ_{N_2}[2]	Φ_{N_2} (sens)[3]	Ref.
CH_3	0.15	0.009	5,43,55a)
C_2H_5	0.023	—	56)
n—C_3H_7	0.007	—	5)
i—C_3H_7	0.025	0.001[4]	5,44)
t—C_4H_9	0.46	0.016	46,55)
$C_6H_5CMe_2$	0.36	0.31	5)
$NCCMe_2$	0.44	0.11	5,57)
$H_2C{=}CHCMe_2$	0.57	0.54	5,58)
$HC{\equiv}CCMe_2$	0.47	0.27	5,58)
$EtOOCCMe_2$	0.42	0.06	59a,b)
Ph and CH_3	0.051	—	52)
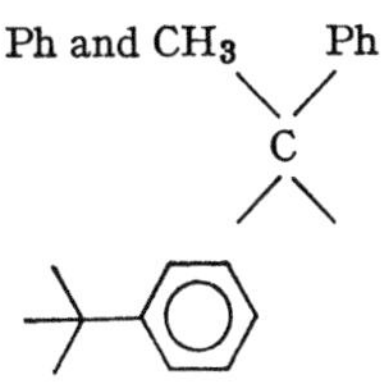	0.58	0.42	60)

[1]) All compounds are *trans* isomers.
[2]) Direct irradiation.
[3]) Benzophenone sensitized.
[4]) Quantum yield for hydrocarbon formation.

3.2.1. Acyclic Azo Compounds

Three typical examples of the photodecomposition are presented in the following scheme:

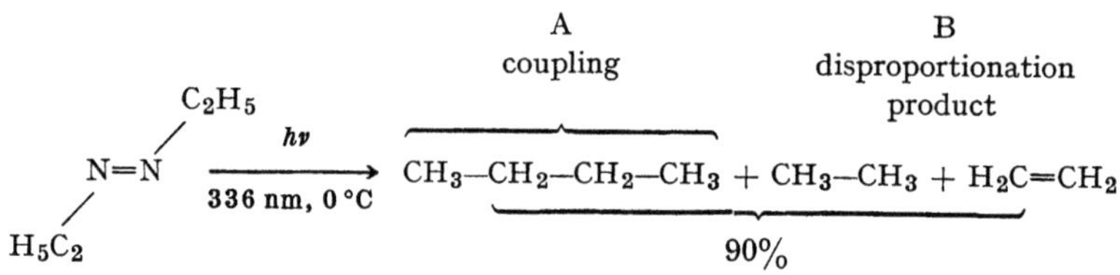

56)

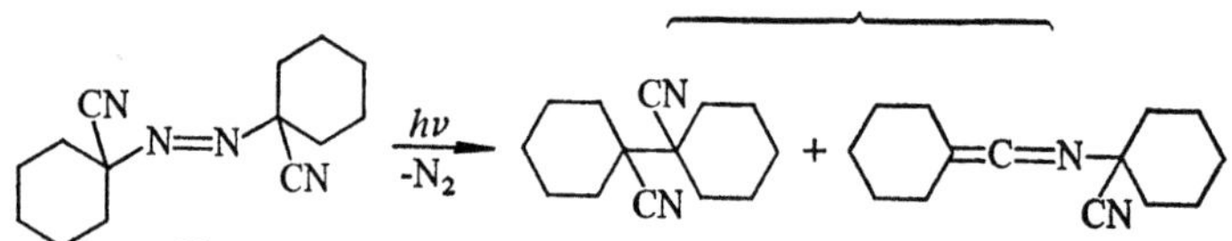

59b)

39 *40* *41*

	40	
dir.:	19.9%	cage products
sens. (triphenylene):	17.7%	
Δ (80%)	9.4%	

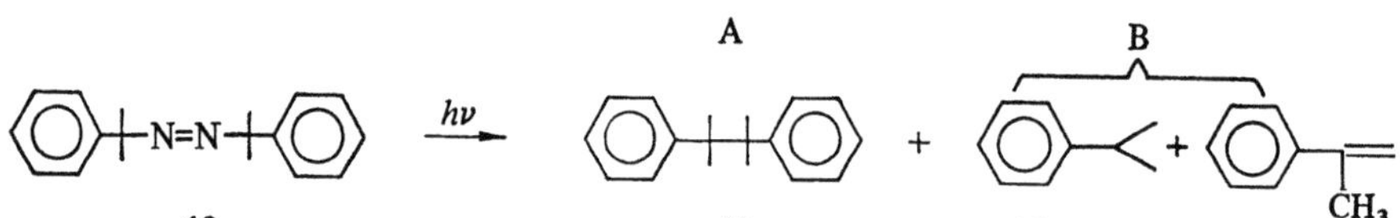

42 *43* *44* *45*

dir: 94–95% 5–6%

60a)

In all reactions nitrogen elimination occurs to give the radicals according to the scheme on p. 71. The radicals thus generated can combine to give the coupling product A or form disproportionation products B as usual in radical chemistry. In some of these reactions studied, the *cage effect* was determined. This cage effect is attributed to the products formed in the solvent cage before diffusion and arbitrary reactions of the radicals can occur. For the reactions of azoethane and *42* cage effects of the order of 83% and 33% have been reported. The triplet reaction of *42* shows the same cage effect as the direct reaction [60]. Photolysis of *39* gives the coupling products *40* and *41*. If cumene is used as solvent the formation of *41* is suppressed and all *40* formed is due to the cage products. The cage effects for the S_1 and the T_1 reaction are given by the yield of *40*, which is of the order of 18—19% in both cases [59]. This result is most easily explained by assuming that intersystem crossing is faster than dissociation into the radical pair.

For the triplet *39* a tight radical pair was originally postulated [39]. The photoelimination of nitrogen from *42* has been mainly studied in order to determine the cage effect of the reaction. Direct (S_1) as well as sensitized photolysis (T_1-state) show a cage effect of about 33% versus 27% for the thermal reaction. At lower temperature disproportionation competes more favorably with coupling to *43* [60]. A clear cut relation between the overall rate constant of azo compound photolysis and cage recombination parameters is however not operative [61]. The sensitized photolysis of other aromatic azo compounds has not yet been investigated, only the direct photolysis *via* the S_1 state has been reported [52,62]. From Table 6 it can be seen that the quantum yields for photodecomposition of acyclic azo compounds (usually in hydrocarbon solvents) are about 50%. In all cases the quantum yields for the triplet reaction are lower, sometimes more than by a factor of 10. This suggests, if energy transfer conditions were met, that the triplet state decomposes to a much lesser extent than does the corresponding singlet state. The only exceptions are azo compounds bearing π-systems on the α-carbon atom of the azo group. If the energy transfer has been complete, decomposition of the triplet is obviously as fast as for the singlet. Nitrogen elimination is therefore most easily explained by assuming the following reaction sequence:

Energy transfer populates the triplet state ($^3\pi\pi$ state) of the azo compound. Then *cis-trans* isomerisation occurs leading to the *cis*-isomer, which is in most cases unstable and decomposes thermally at room temperature.

In the gas phase however decomposition can occur with unit efficiency [1]. Here electronically as well as vibrationally excited molecules are involved in the decomposition step. The azomethane photolysis in the gas phase leads to decomposition with $\phi_{N_2} = 1$ at lower pressures [55b]. With increasing pressure the quantum yield ϕ_{N_2} drops from 1 to lower values. This is attributed to partial intersystem crossing of the excited azoalkane to the triplet state which gives only *cis/trans* isomerisation.

3.2.2. Nitrogen Elimination from Cyclic Azo Compounds

The photochemical extrusion of nitrogen from cyclic azo compounds has been studied on a broad scale. Cyclic azo compounds, in contrast to acyclic ones can react with N_2 elimination from both the S_1 and the T_1 state. Ring strain as well as the lack of *cis/trans* isomerisation favor photodecomposition. Since intersystem crossing is not very efficient the T_1 state has mostly been populated by energy transfer.

The lowest members of cyclic azocompounds are the *diazirines*. Photofragmentation of diazirines affords carbenes which can react in the normal way [63].

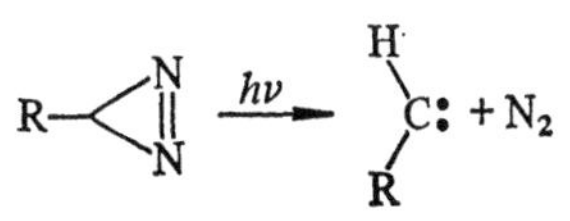

Most of these reactions have been carried out in the gas phase. At higher pressure triplet carbenes are involved. However it is not clear if intersystem crossing occurrs already in the diazirine or in the carbene formed [63].

The photolysis of pyrazolines has attracted much attention since in this reaction 1,3-diradicals are produced, which were supposed to be related to the intermediate trimethylene diradical which would be a model for an addition of singlet carbenes to olefins. The sensitized decomposition should resemble the trimethylene diradical of the triplet carbene addition to double bonds. The stereochemical course of pyrazoline photolysis can be determined by using disubstituted pyrazolines. Thus photolysis of the dimethyl-pyrazolines *46*–*49* affords the following products [64]:

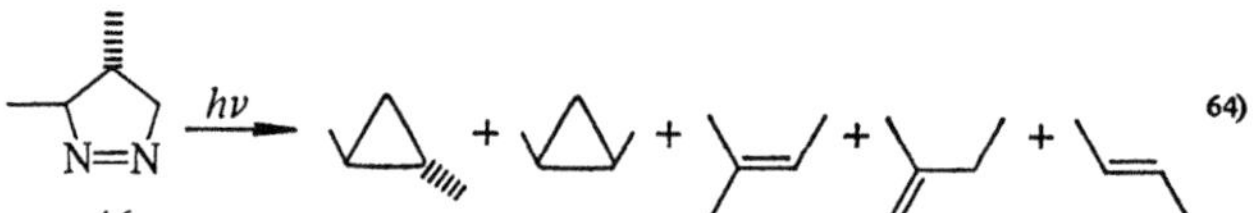

46 [64]

Δ:	21.8	46.0	16.3	15.8	–
hν gas phase:	37.2	24.4	16.4	14.5	5.7
EtOH:	50.7	13.0	18.3	14.1	3.1
sens: Ph_2CO, EtOH:	72.7	27.8	1.5	1.1	1.9

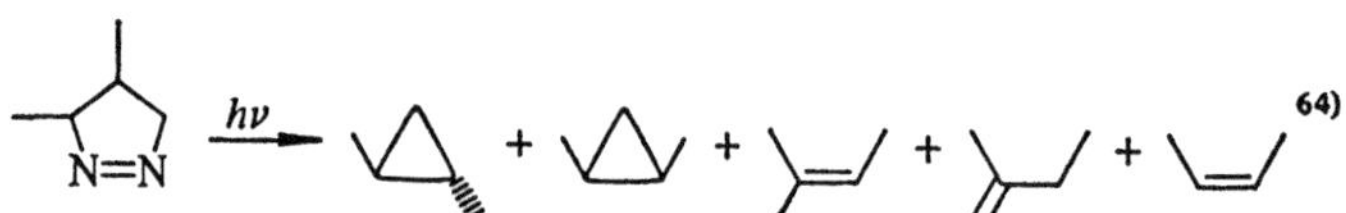

47 [64]

Δ:	35.2	43.7	14.4	7.1	–
hν: gas phase:	37.6	43.0	4.7	4.8	7.2
EtOH:	26.4	60.4	5.0	3.7	4.0
sens: EtOH:	66.2	32.1	0.5	–	1.3

48 [64]

direct (EtOH):	45.6	44.5	2.8	3.9	2.8
sens (Ph_2CO):	61.2	38.8	trace	trace	trace

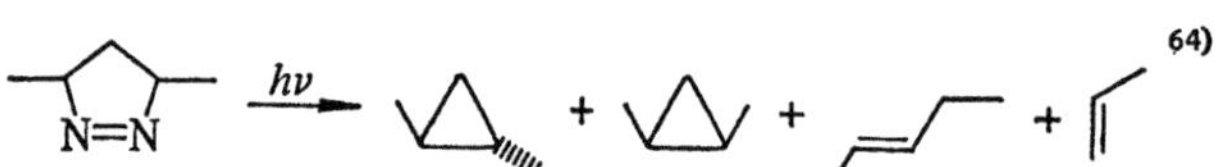

49 [64]

direct (EtOH):	61.7	33.9	1.0	2.7
sens (Ph_2CO):	60.1	38.9	<0.2	0.8

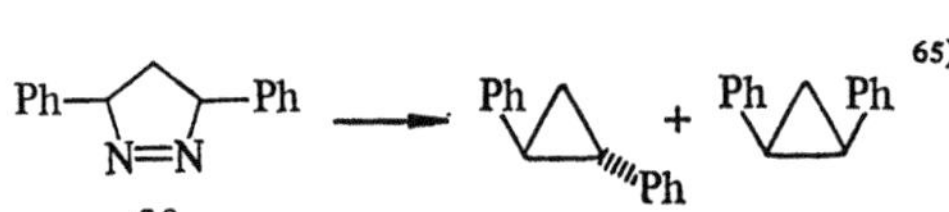

50 [65]

51

	A	*B*	*A*/*B*
direct:	88	12	7:1
sens:	65	35	1.8:1

For easier comparison the result of the thermal reaction is included for compounds *46* and *47*. As can be seen in the reaction scheme above direct photolysis of the pyrazolines *46* and *47* proceeds mainly with retention of the original stereochemistry in the cyclopropanes isolated. *48, 49* and *50* however lead mainly to the inverted stereochemistry in the cyclopropanes. The singlet biradical *51* formed from *46—49* is therefore clearly not on the same energy surface as a „possible" singlet diradical in the carbene cycloaddition. However one knows today that singlet carbene cycloaddition is a concerted process, so no such diradical should be detectable.

The 0,0-diradical or π-cyclopropane *51* is also unlikely as thermal decomposition reactions show (see [86]).

In the triplet reaction however, the stereochemistry is lost and an almost similiar ratio of *cis/trans*-cyclopropanes is obtained from *46, 47* and *48, 49*. The *trans/cis*-cyclopropane ratio for *46* of 2.77 and *47* of 2.63 is in good agreement with the ratio of 2.9 found for the triplet addition of :CHT to *cis-* or *trans-* but-2-ene [66]. This shows that rotation around the C—C-bond in the diradical *51* is faster than spin inversion.

The results for *46—50* clearly show the spin correlation effect for the different electronic states involved. Shorter wavelengths of irradiation can enhance the spin correlation effect in pyrazolines [67].

A clear spin correlation effect is observed for the pyrazolines *4a* and *4b* [68,69]:

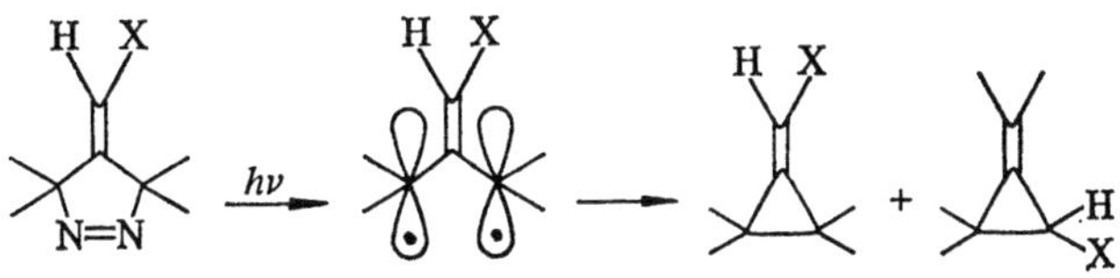

4a; X = Cl	direct:	77	23 [68]
	sens. (Ph_2CO):	25	75
4b; X = CO_2R	direct:	64	36
	sens.:	18	82

For the 6-membered ring azo compound *52* spin correlation effect is operative, too [70].

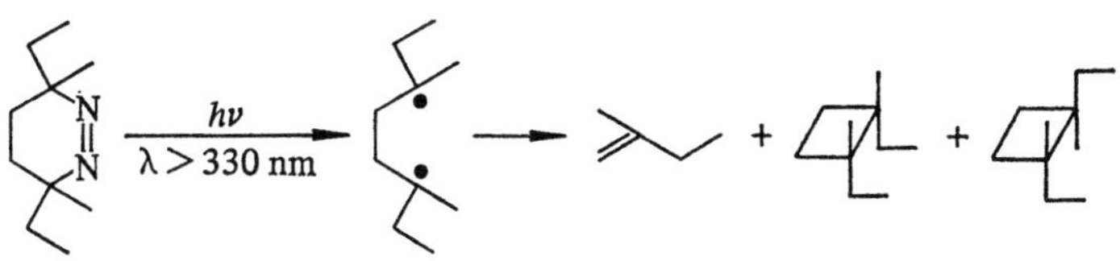

52	*53*	*54*	cis-*55*	trans-*55*
meso-*52*	direct:	61%	35%	3.5%
d,l-52		60	4	33
meso-*52*	sens. (thio-	77	11.5	8
d,l-52	xanthone):	75	8	12

Formation of *cis*- and *trans*- cyclobutanes *55* occurs with 95–97% retention of configuration. The triplet reaction however gives only 60–65% retention of configuration. In the triplet biradical *53* the parallel spins delay ring closure thus allowing rotations about all three single bonds. The amount of cleavage to give *54* is also enhanced in the triplet reaction of *52*. The retention of configuration in the direct photolysis of *52* is in contrast to the pyrazoline photodecomposition of *48* and *49* where mainly inversion of stereochemistry occurs (see above). Spin correlation effects are also reported for 7- and 8-membered ring azo compounds [71].

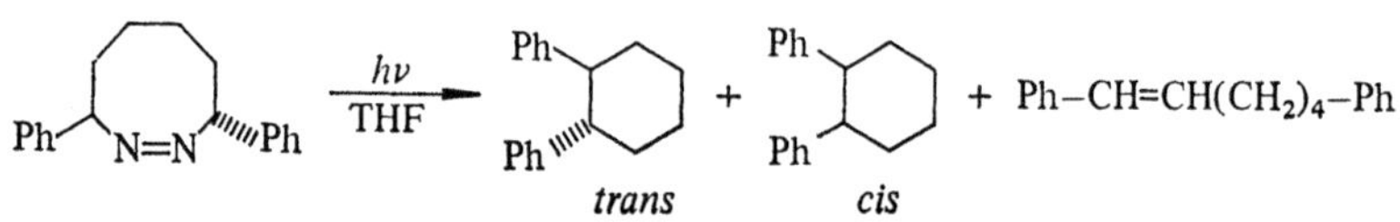

	trans	cis	
direct:	20%	15%	65%
sens. (Ph_2CO):	traces		>95%

3H-Pyrazoles: Azo compounds of the 3*H*-pyrazole type are an interesting class of molecules. The basic mechanism for the photochemical decomposition of cyclic azo compounds becomes more complex for 3*H*-pyrazoles because of the additional double bond present.

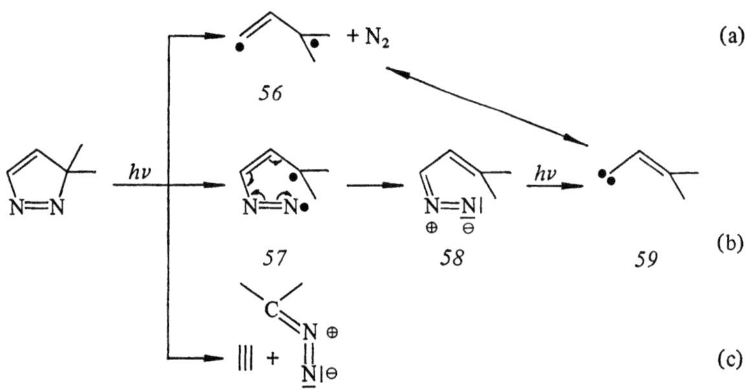

In addition to the geminate and free radical pairs a 1,3-retro-dipolar path (c) is possible in certain systems. The concerted scission of two bonds (a) leads to a vinyl-allyl-diradical. One-bond scission produces the diradical *57* which can rearrange to the vinyl diazoalkane *58*. A second photon can then give rise to the vinyl carbene *59* which is one of the resonance forms of diradical *56*.

A series of substituted 3*H*-pyrazoles has been irradiated since this reaction is an easy entry to the class of cyclopropenes [72–76].

No sensitization experiments to generate the excited triplet states of 3*H*-pyrazoles have been reported so far. On the other hand the triplet multiplicity of the intermediates has been detected by ESR.

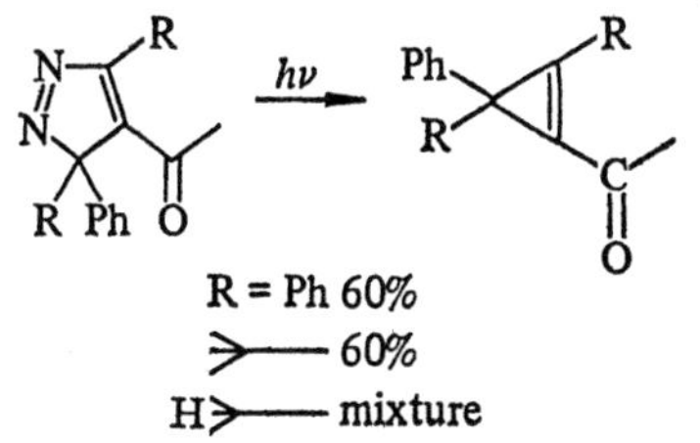

[73d]

Photolysis of the 3*H*-pyrazole *60* with a Cu/Ni-salt filter (λ: 330—410 nm) afforded the vinyldiazoalkane *61*. This could be interconverted to the cyclopropene *63* *via* the vinylcarbene *62*. Direct photolysis with light $\lambda > 350$ nm gives *63* directly. ESR spectroscopy after irradiation at 5 K indicated clearly a triplet carbene *62* [76].

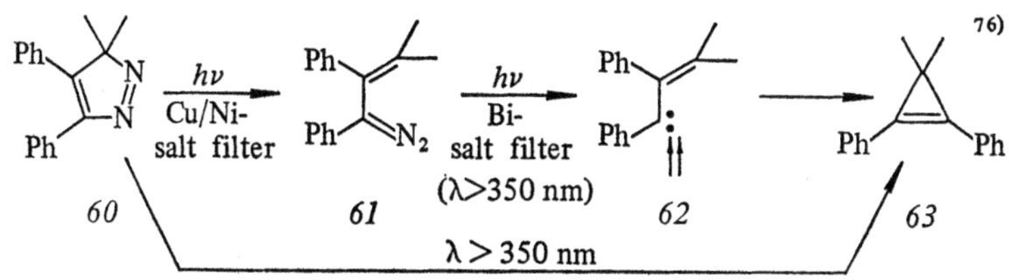

[76]

An unexpected strange reaction of 3*H*-pyrazole *64* giving *65* *via* a carbene dimerisation followed by [2+2]-cycloaddition and *66* *via* a hydrogen abstraction followed by cyclisation has been reported. No clear statement was made on the multiplicity of that reaction [77].

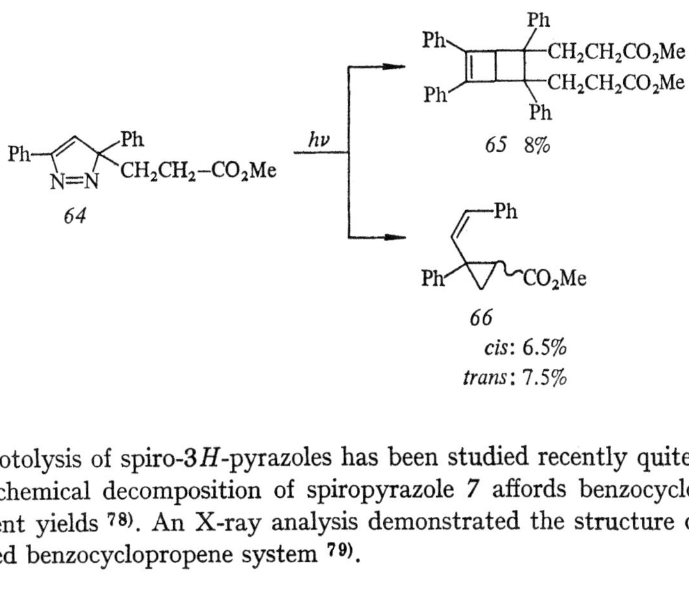

Photolysis of spiro-3*H*-pyrazoles has been studied recently quite intensively. Photochemical decomposition of spiropyrazole *7* affords benzocyclopropenes in excellent yields [78]. An X-ray analysis demonstrated the structure of the highly strained benzocyclopropene system [79].

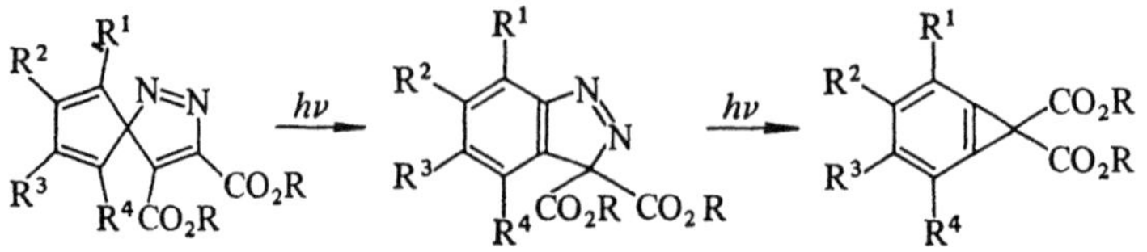

7 a–d, f *67 a–d, f* *68 a–d, f*

	R^1	R^2	R^3	R^4	%
a	Ph	Ph	Ph	Ph	85
b	Ph	H	H	Ph	86
c	Ph	Ph	Ph	H	26
d	Ph	Ph	benzo		30
e	Ph	m–Cl–C_6H_4	benzo		10
f	Ph	p–Cl–C_6H_4	benzo		–

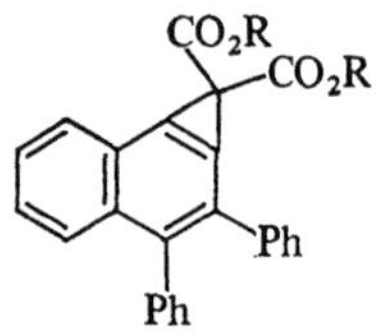

68d

% *68a*
direct: 87%
sens: 85%
$(CH_3)_2CO)$

The first naphtho[a]cyclopropene *68d* was also obtained by that route. 3*H*-indazole was proposed as an intermediate in this reaction [78]. The reaction proceeds equally well from the triplet state of 7. However with acetone light capture from the sensitizer may not have been complete. A trapping of the diradical intermediate *67'* to give the known indene *69* by photolysing *7* in dimethyl-acetylene-dicarboxylate (ADC) afforded only a trace of *69*. However the yield of *68a* was reduced to 40%. Either trapping of *67'* was not effective enough or ADC may have acted as quencher. Trapping with furan or cyclopentadiene was also not effective [78a].

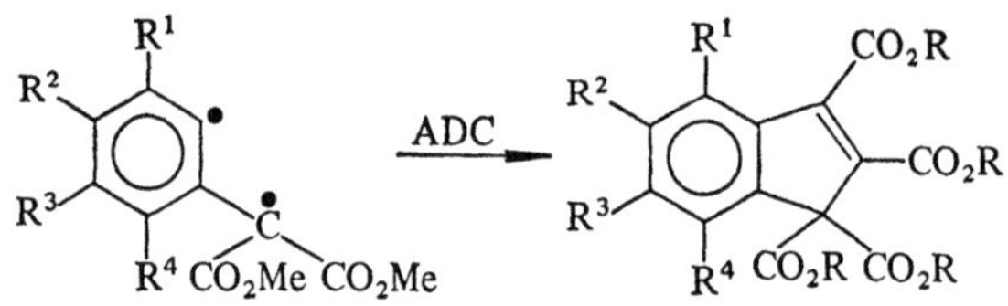

67' *69*

R=CH_3

In the meantime the first spiro-benzocyclopropenes *68g* could be prepared [78b].

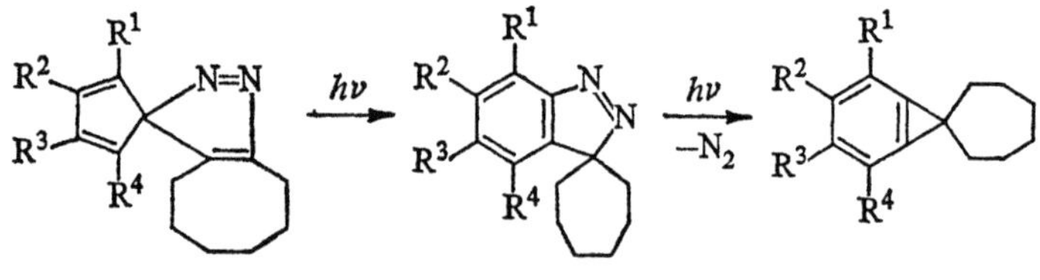

7g *67g* *68g*

The separate irradiation of 3*H*-indazoles *67* was less effective in N_2-evolution in the sensitized as in the direct run [37]. With piperylene also no quenching was observed for *67*. ESR has shown that the ground state of the biradical produced by photolysis of *67* is a triplet [36]. So it was concluded that after photolysis of *67* a diradical *67'* is generated in the singlet state which either reacts to *68* or undergoes intersystem crossing to triplet *67'*. The photodecomposition of dibenzo-3*H*-pyrazoles *70* gave rise to [1.2]spirenes *73* [80,81a)].

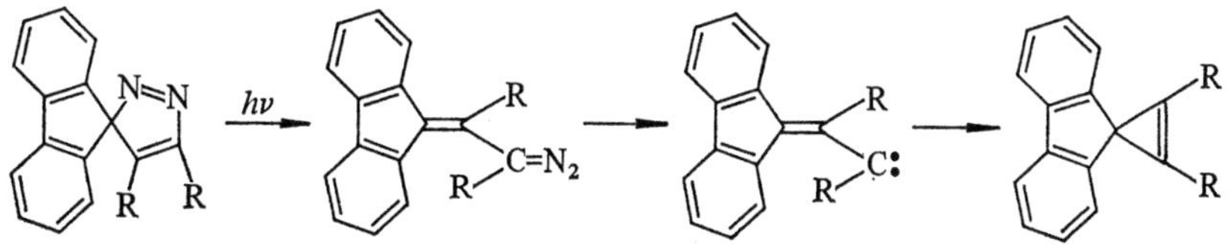

70 *71* *72* *73*.

R = CH_3CO; Ph CO; CF_3

Quenching experiments (anthracene) as well as sensitisation (xanthone) show that in that reaction a triplet is involved [78b)]. This is not clear in the photolysis of *A*, which leads also to a cyclopropene derivative *B* [78c)].

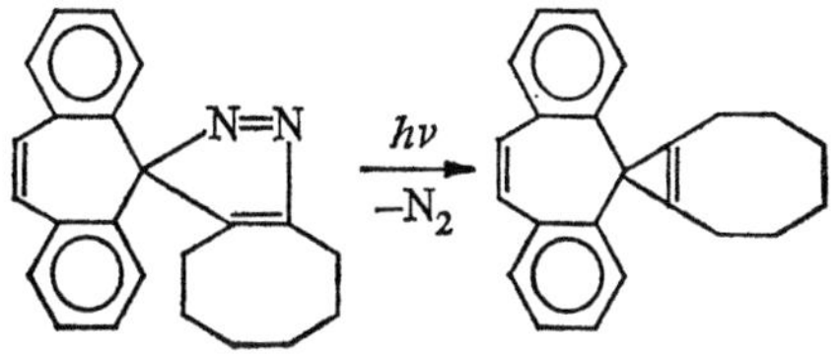

A *B*

The direct photolysis leads after ring opening to vinyldiazocompound *71* which decomposes *via* the carbene *72* to *73*. Mechanistic studies demonstrated that the rather selective carbene *72* involved adds nonstereospecifically to *cis*- or *trans*-4-methyl-pent-2-ene and therefore from a triplet state [81)]. The difference of the reaction path for the photolysis of 3*H*-pyrazoles *74* leading to *68* d, *e* (R'=H, m—Cl) or [1.2]-spirene *75* (R'=p—Cl) may be attributed to intersystem crossing induced

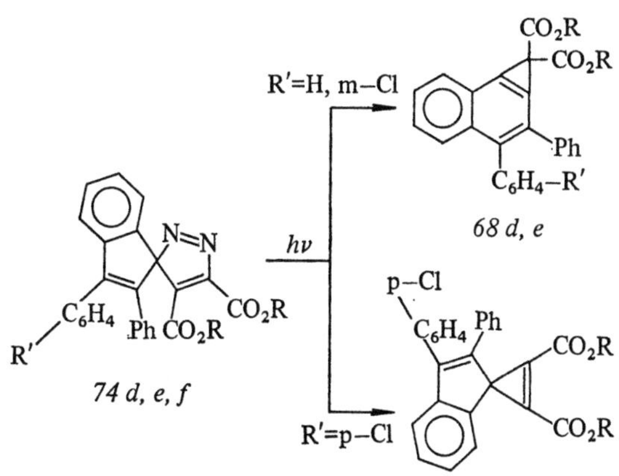

68 d, e

74 d, e, f

R=CH_3

75 f : 61%

by the chlorine heavy atom effect being effective only in the p-position. This result would thus indicate a spin correlation effect [82]. However more work needs to be done with this problem to clarify this point unambiguously.

3.2.3. Nitrogen Elimination from Bicyclic and Polycyclic Azo Compounds

Bicyclic azo compounds can decompose photochemically in a way similiar to monocyclic ones, however here 1,3-retrodipolar fragmentation can occur also. So 2,3-diazabicyclo [3.1.0]hex-2-ene *76* can photofragment *via* the allyl-diazoalkane *77* and the allylcarbene *78* to *79* and *80* [83].

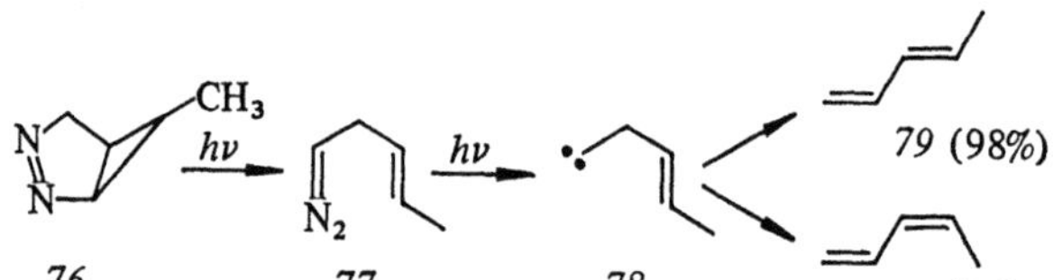

76 *77* *78* *79* (98%) *80* (2%)

yield: *79* + *80* = 86%

The 4,4-diphenylderivative of *76* however showed coupling products of a 1,3-diradical in addition [84]. This reaction was however only studied for the singlet species. This study implied a one-bond scission as the major step in the reaction. The homolog *81* however showed equal product ratios for the photodecomposition in direct as well as sensitized reactions [85] (yields refer to degree of stereospecifity).

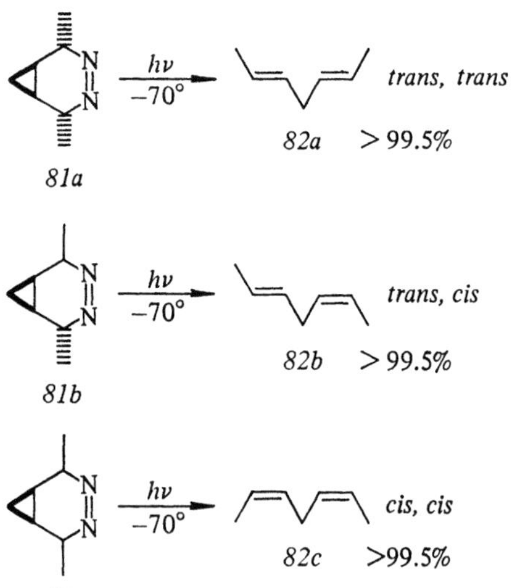

81a → *82a* > 99.5%

81b → *82b* > 99.5%

81c → *82c* >99.5%

Despite the fact that photolysis of *81a* can be sensitized, no quenching occurs on the other hand for the direct photolysis of *81b*. The photochemical reaction is thought to occur in a disrotatory fashion leading to the dienes *82*. For the thermal reaction, on the other hand, a highly twisted intermediate would result because of the conrotatory ring opening leading to a *trans* azo group. This reasoning holds

81a

good however only for a concerted ring opening. Since most nitrogen eliminations proceed *via* a one-bond scission [86], these explanations seem somewhat in doubt (see above).

A spin correlation effect was also noted in *83* [87],

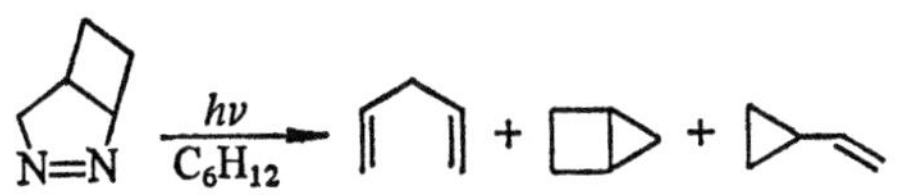

83 *84a*

direct:	26.8	58.9	10.7
sens. (acetophenone):	10.3	89.7	trace

as well as in the photolysis of diaza-bicyclo[3.3.0]oct-3-ene [88].

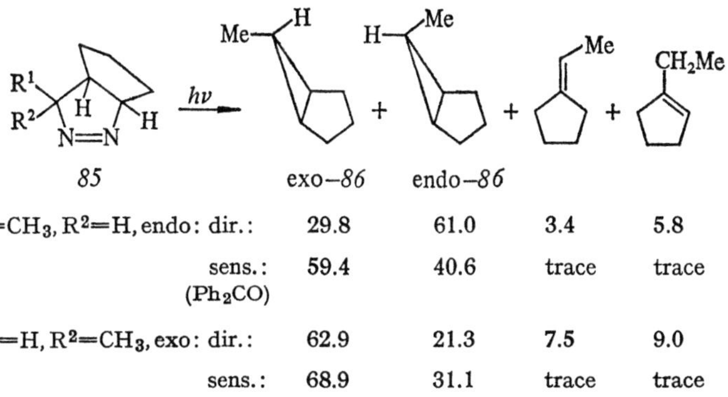

85 exo–*86* endo–*86*

$R^1=CH_3, R^2=H$, endo:	dir.:	29.8	61.0	3.4	5.8
	sens.: (Ph_2CO)	59.4	40.6	trace	trace
$R^1=H, R^2=CH_3$, exo:	dir.:	62.9	21.3	7.5	9.0
	sens.:	68.9	31.1	trace	trace

In the singlet reaction exo-*86* and endo-*86* are produced with predominant retention of configuration. The triplet reaction shows an almost identical product distribution starting from exo-*85* or endo-*85*. One-bond scission and slower rate of ring closure of the triplet biradical explain these facts sufficiently well. A similar spin correlation effect was found for the hetero analog of *85* [89].

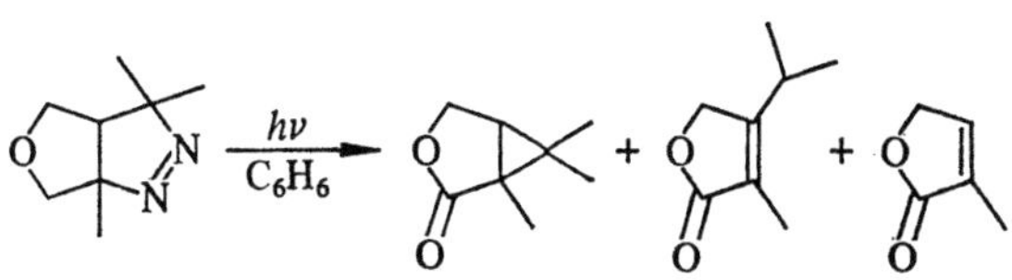

85'

direct:	62%	21%	17%
sens. (Ph_2CO):	99%	—	—

Much work has been done on the photodecomposition of 2,3-diaza-bicyclo-[2.2.1]hept-2-ene *3* [17,23].

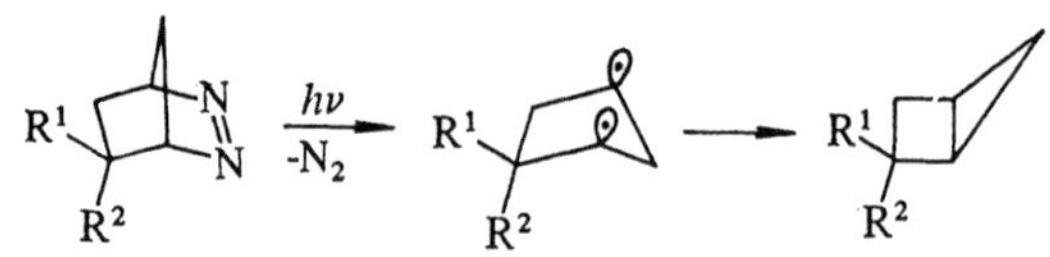

3a-c *84a-c* : exo or endo

a: R^1=H, R^2=H; b: R^1=CH_3, R^2=H; *c*: R^1=H, R^2=CH_3

Singlet as well as triplet photodecomposition has been reported.

The 5-methoxy derivatives of *3b* or *3c* show a clear spin correlation effect. The singlet reaction gives different product ratios for exo- and endo-*84*, whereas the triplet reaction gives the same ratio 79/21 for both stereoisomers. From this result an equilibrated pyramidal or planar diradical was inferred as an intermediate [90].

Nitrogen elimination from the triplet state was also reported for *8* [27]. No sensitized reactions were carried out for the azo compounds described in *l.c.*[91].

Photolysis as well as thermolysis of *A* generate a triplet biradical which was detected by ESR [92].

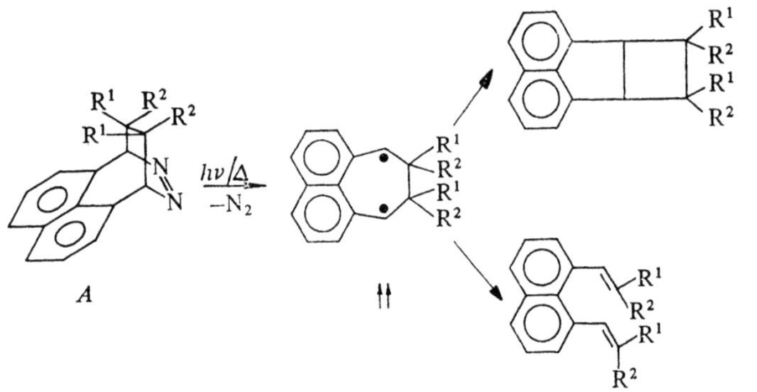

A spin correlation effect is found also for the pentacyclic system *87* [93].

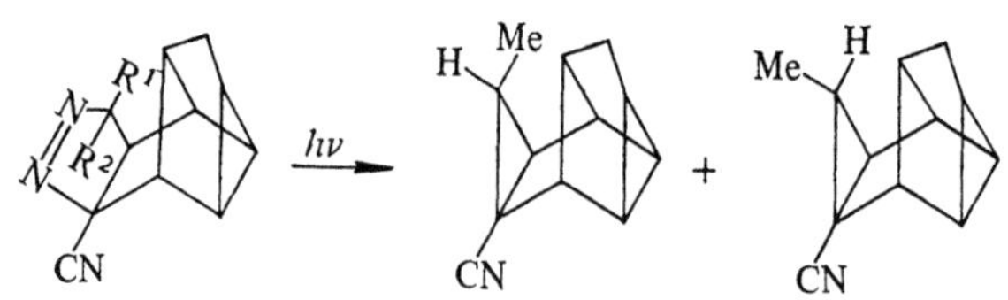

87 endo–*88* exo–*88*

		endo–*88*	exo–*88*
R^1=H, R^2=Me(exo):	dir.:	60.5	7
	sens.: (Ph_2CO)	93	7
R^1=CH_3, R^2=H(endo):	dir.:	22	51
	sens.:	63	37

Table 7. Nitrogen quantum yields for direct photolysis of cyclic azo compounds in nonpolar solvents

Compound	Φ_{N_2}	Ref.
	(0.7–1.8; gasphase)	67)
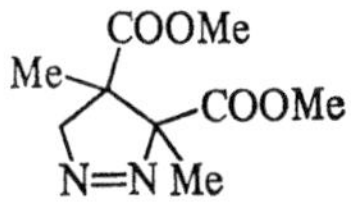	0.52	5)
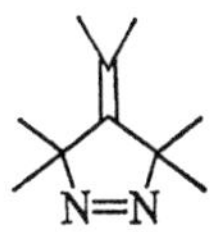	0.45	5)
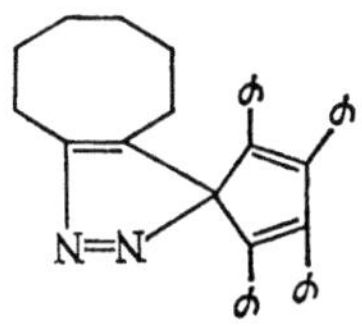	0.026	78b)
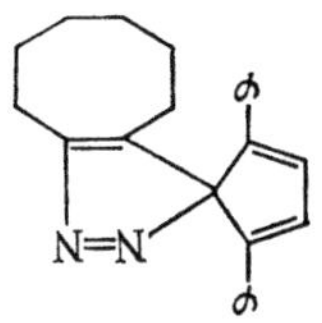	0.025	78b)
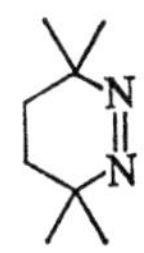	0.002	5)
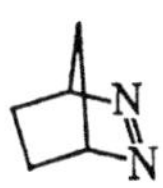 *3*	1.0	
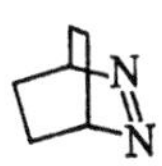 *8a*	0.022	27)
 *8b*	0.014	5)

The singlet leads to the configurationally inverted forms, whereas the triplet mainly affords endo-*88*. This result indicates that rotation is inhibited in the diazenyl radical formed primarily. In the triplet diradical the formed ring closure is obviously slower than in the singlet diradical.

In Table 7 the quantum yields Φ_{N_2} for photolysis of cyclic azo compounds are presented. So far only one determination of a triplet Φ_{N_2} has been published.

The energy gap between the S_1 and the T_1 state in cyclic azo compounds depends obviously on ring size [5]. The larger this energy gap the less probable is intersystem crossing to the triplet. *8*, in contrast to *3*, undergoes intersystem crossing as well as emission. Only in very few cases have intersystem crossing efficiencies been measured. All examples presented in this chapter show a clear *spin correlation effect*. The exceptions are however the azo compounds *49* and *81*, which show the same product distribution in singlet and triplet reactions. This could be due (a) *to inefficient sensitization*, (b) *reaction from an equilibrating state* (c) *from a* S_0 *ground state reaction* or (d) *fast intersystem crossing steps*. In *acyclic azocompounds no clear* spin correlation effect has been observed. This is due to (a) singlet energy transfer or (b) thermal decomposition of photochemically generated *cis*-azo compounds.

4. Miscellaneous Reactions

Azo compounds bearing α-hydrogen atoms can photochemically, as well as thermally, tautomerize to the more stable hydrazones. Little has been done however in studying the multiplicity effect of this reaction. Phototautomerisation was reported for the following compounds:

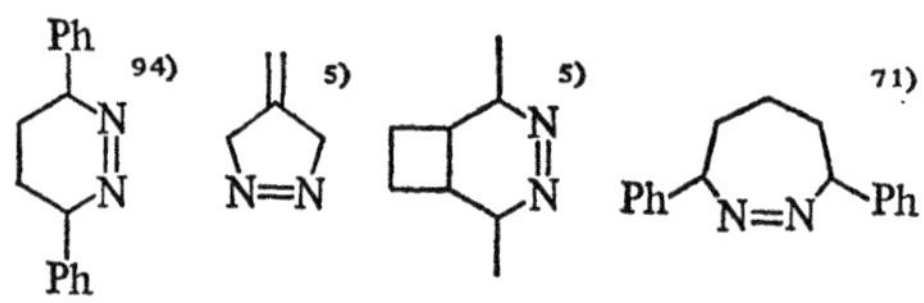

Another possible route is photoreduction from *89* to hydrazinderivatives [95].

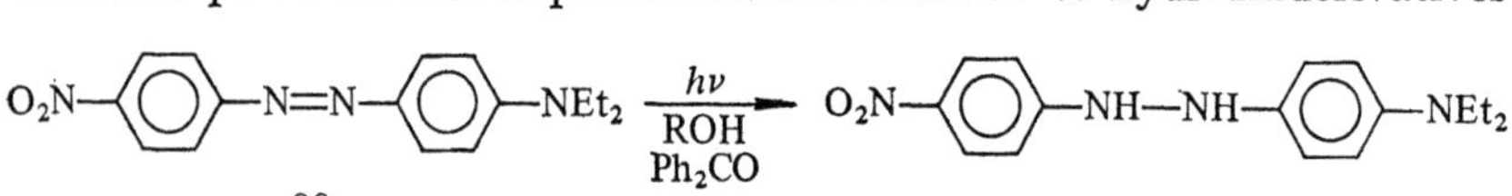

89

So azo-benzene derivatives can be photoreduced — most probably *via* a triplet state — to hydrazo compounds. In some aromatic azo compounds photocyclisation occurs also probably *via* a triplet state, protonation can increase the yield in these reactions [96].

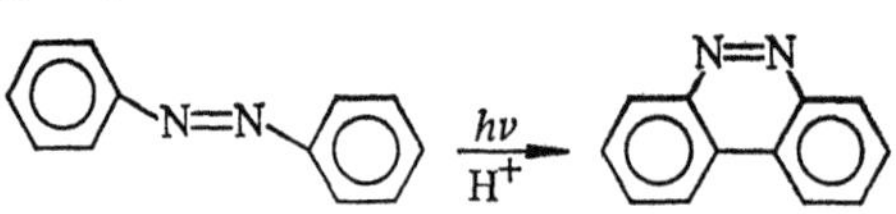

$\Phi = 0.015$

We take the opportunity here to express our thanks to the Deutsche Forschungsgemeinschaft and to the Fonds der Chemischen Industrie for their support of this present work.

Our thanks go also to Dr. Ranade for his checking and Miss Halberstadt for her typing of the manuscript.

5. References

[1] Calvert, J. G., Pitts, J. N., Jr.: Photochemistry, p. 462. New York: John Wiley and Sons, Inc. 1966.
[2a] Wagner, P. J., Hammond, G. S.: Advances in photochemistry, Vol. 5, p. 130. New York: Interscience Publications 1968.
[2b] Turro, N. J., Dalton, J. C., Weiss, D. S.: Organic photochemistry (O. Chapman, M. Dekker, Eds.) New York: 1969.
[3] Bryce-Smith, D.: Special periodical reports. Chem. Soc. (London) *1*, 402 (1970); *2*, 709 (1971) and *3*, 758 (1972).
[4] Griffiths, J.: Chem. Soc. Rev. *1*, 481 (1972).
[5] Engel, P. S., Steel, C.: Accounts Chem. Res. *6*, 275 (1973).
[6] Rau, H.: Angew. Chem. *85*, 248 (1973).
[6a] Gleiter, R., Dürr, H.: (unpublished).
[7] Stern, E. S., Timmons, C. J.: Electronic absorption spec. in org. chem., London: E. Arnold 1970.
[8] Dürr, H., Schrader, L.: Z. Naturforsch. *24b*, *5*, 536 (1969).
[9] Dürr, H., Schrader, L.: Chem. Ber. *103*, 1334 (1970).
[10] Suzuki, H.: Electronic absorption spectra and geometry of organic molecules, p. 499. New York: Academic Press 1967.
[11] Kearns, D. R.: J. Phys. Chem. *69*, 1062 (1965). Gordon, M. S., Fischer, H.: J. Am. Chem. Soc. *90*, 2471 (1968).
[12] Robin, M. B., Hart, R. R., Kuebler, N. A.: J. Am. Chem. Soc. *89*, 1564 (1967).
[13] Hochstrasser R. M., Lower, S. K.: J. Chem. Phys. *36*, 3505 (1962).
[14] Chen, S. Y., Hedges, R. M.: Theoret. Chim. Acta *31*, 275 (1973).
[15] Baird, N. C., De Mayo, P., Swenson, J. R., Usselmann, M. C.: Chem. Commun. *1973*, 314.
[16] Steel, C., Thomas, T. F.: Chem. Commun. *1966*, 600.
[17] Solomon, B. S., Thomas, T. F., Steel, C.: J. Am. Chem. Soc. *90*, 2249 (1968).
[18] Rau, H.: Luminescence *1/2*, 191 (1970).
[19] Cohen, B. J., Baba, H., Goodman, L.: J. Chem. Phys. *43*, 2902 (1965).
[20] Müller, R., Dörr, F.: Ber. Bunsenges. Physik. Chem. *63*, 1150 (1959).
[21] Rau, H.: Ber. Bunsenges. Physik. Chem. *72*, 408 (1968).
[22] Dürr, H., Rauh, H.: (unpublished), a series of derivatives of *7* show phosphorescence as well as fluorescence.
[23] Engel, P. S.: J. Am. Chem. Soc. *89*, 5731 (1967); *91*, 6903 (1969).
[24] Fischer, E.: J. Am. Chem. Soc. *90*, 796 (1968).
[25] Hochstrasser, R. M., Marzacco, C.: J. Phys. Chem. *46*, 4155 (1967).
[26] Collier, S. S., Slater, D. H., Calvert, J. G.: Photochem. Photobiol. *7*, 737 (1968).
[27] Clark, W. D., Steel, C.: J. Am. Chem. Soc. *93*, 6347 (1971).
[28] Jones, L. B., Hammond, G. S.: J. Am. Chem. Soc. *87*, 4219 (1965).
[29] Wamser, C. C.: J. Am. Chem. Soc. *95*, 2044 (1973).
[30] Lower, S. K., El-Sayed, M. A.: Chem. Rev. *66*, 199 (1966); Accounts Chem. Res. *1*, 8 (1968).
[31] Haselbach, E., Heilbronner, E.: Helv. Chim. Acta *53*, 684 (1970); Haselbach, E., Schmelzer, A.: Helv. Chim. Acta *54*, 1575 (1971).
[32] Boyd, R. J., Bunzli, J. C., Snyder, J. P., Heymann, M. L.: J. Am. Chem. Soc. *95*, 6479 (1974).
[33] Robin, M. B., Basch, H., Kuebler, N. A., Wiberg, K., Ellison, G. B.: J. Chem. Phys. *51*, 45 (1969).
[34] Marnet, L. J., Smith, P., Porter, N. A.: Tetrahedron Letters. *1973*, 1081. For similiar experiments see also: Brumby, S.: J. Magn. Res. *10*, 203 (1973). — Stilbs, P., Ahlgren, G., Akermark, B.: Tetrahedron Letters *1972*, 2387.
[35] Bartlett, P. D., McBride, J. M.: Chem. Eng. News *44*, *41*, 106 (1966).
[36] Closs, G. L., Kaplan, L. R.: J. Am. Chem. Soc. *91*, 2168 (1969). Gloss, G. L., Kaplan, L. R., Bendall, V. I.: J. Am. Chem. Soc. *89*, 3376 (1967).
[37] Schrader, L.: Tetrahedron *29*, 1833 (1973).

[38] Roth, H. P.: J. Am. Chem. Soc. *93*, 4935 (1971).
[39] Porter, N. A., Marnett, L. J., Lochmüller, C. H., Closs, G. L., Shobataki, M.: J. Am. Chem. Soc. *94*, 3665 (1972).
[40] Closs, G. L., Trifunac, A. D.: J. Am. Chem. Soc. *91*, 4554 (1969).
[41] Mill, T., Stringham, R. S.: Tetrahedron Letters *1969*, 1863.
[42] Alster, J., Burnelle, L. A.: J. Am. Chem. Soc. *89*, 1261 (1967).
[43] Hutton, R. F., Steel, C.: J. Am. Chem. Soc. *86*, 745 (1964).
[44] Abram, J. J., Milne, G. S., Solomon, B. S., Steel, C.: J. Am. Chem. Soc. *91*, 1220 (1969).
[45] Kaldor, Y., Shavitt, J.: J. Chem. Phys. *48*, 191 (1968).
[46] Bartlett, P. D., Engel, P. S.: J. Am. Chem. Soc. *90*, 2960 (1968).
[47] Fischer, E., Frankel, M., Wolovsky, R.: J. Chem. Phys. *23*, 1367 (1955).
[48] Zimmerman, G., Chow, L., Paik, V. J.: J. Am. Chem. Soc. *80*, 3528 (1958).
[49] Talaty, E. R., Fargo, J. C.: Chem. Commun. *1967*, 65. Schulte-Frohlinde, D.: Liebigs Ann. Chem. *612*, 138 (1958).
[50] Yamashita, S., Ohno, H., Tayama, O.: Bull. Chem. Soc. Japan *35*, 1849 (1962).
[51] Dyck, R. H., McClure, D. S.: J. Chem. Phys. *36*, 2326 (1962).
[52] Porter, N. A., Marett, L. J.: J. Am. Chem. Soc. *95*, 4361 (1973).
[53] Ramsperger, H. C.: J. Am. Chem. Soc. *51*, 2134 (1929).
[54] Seltzer, S., Dunne, F. T.: J. Am. Chem. Soc. *87*, 2628 (1965).
Pryor, W. A., Smith, K.: J. Am. Chem. Soc. *92*, 5403 (1970). Tagaki, K., Crawford, R. J.: J. Am. Chem. Soc. *93*, 5910 (1971).
[55a] Engel. P. S., Bartlett, P. D.: J. Am. Chem. Soc. *92*, 5883 (1970).
[55b] Rebbert, R. E., Ausloos, P.: J. Am. Chem. Soc. *87*, 1847 (1965).
[56] Kodama, S., Fujita, S., Takeishi, J., Toyama, O.: Bull. Chem. Soc. Japan *39*, 1009 (1966). Cerfontain, H., Kutschke, K. O.: Can. J. Chem. *36*, 344 (1958).
[57] Smith, P., Rosenberg, A.: J. Am. Chem. Soc. *81*, 2037 (1959). Back, R., Sivertz, C.: Can. J. Chem. *32*, 1061 (1954).
[58] Engel, P. S., Bishop, D. J.: J. Am. Chem. Soc. *94*, 2148 (1972).
[59] Hammond, G. S., Fox, J. R.: J. Am. Chem. Soc. *86*, 1918, 4031 (1964).
[60] Nelson, S. T., Bartlett, P. D.: J. Am. Chem. Soc. *88*, 143 (1966).
[61] Weiner, S. A., Hammond, G. S.: J. Am. Chem. Soc. *91*, 986 (1969).
[62] Horner, L., Neumann, W.: Liebigs Ann. Chem. *587*, 93 (1954). Mackay, D., Marx, U. F., Waters, W. A.: J. Chem. Soc. 1964, 4793. For thermal decomposition see also: Kopecky, K. R., Gillan, T.: Can. J. Chem. *47*, 2371 (1969).
[63] Frey, H. M.: Advances in photochemistry, Vol. 4. New York: Interscience Publications 1966. Amrich, M. J., Bell, J. A.: J. Am. Chem. Soc. *86*, 292 (1964).
[64] Moore, R., Mishra, A., Crawford, R. J.: Can. J. Chem. *46*, 3305 (1968).
[65] Overberger, C. G., Zangero, R. E., Anselme, J. P.: J. Org. Chem. *31*, 2046 (1966). Overberger, C. G., Weinshenker, N., Anselme, J. P.: J. Am. Chem. Soc. *87*, 4119 (1965); *86*, 5364 (1964).
[66] McKnight, C., Lee, P. S., Rowland, F. S.: J. Am. Chem. Soc. *89*, 6802 (1967).
[67] Nowacki, S. D., Do, P. B., Dorer, F. H.: Chem. Commun. *1972*, 273. Looper, G. L., Dorer, F. H.: J. Am. Chem. Soc. *95*, 20 (1973).
[68] Andrews, S. D., Day, A. C.: Chem. Commun. *1966*, 667; see also: Andrews, S. D., Day, A. C., Inwood, R. N.: J. Chem. Soc. *1969*, 2443.
[69] Sanjiki, T., Kato, H., Ohta, M.: Chem. Commun. *1968*, 496; *1969*, 638.
[70] Bartlett, P. D., Porter, N. A.: J. Am. Chem. Soc. *90*, 5317 (1968).
[71] Overberger, C. G., Yaroslavsky, C.: Tetrahedron Letters *1965*, 4395. Overberger, C. G., Stoddard, J. W.: J. Am. Chem. Soc. *92*, 4922 (1970).
[72] Closs, G. L., Böll, W. A., Heyn, H., Dev, V.: J. Am. Chem. Soc. *90*, 173 (1968).
[73] Franck-Neumann, M., Buchecker, C.: Tetrahedron Letters *1969*, 15, 2659; *1973*, 937; *1973*, 2875.
[74] Schrader, L.: Chem. Ber. *104*, 941 (1971). Snatzke, G., Langer, H.: Chem. Ber. *102*, 1865 (1969).
[75] Day, A. C., Inwood, R. N.: J. Chem. Soc. C. *1969*, 1065.
[76] Pincock, J. A., Morchat, R., Arnold, D. R.: J. Am. Chem. Soc. *95*, 7536 (1973); Palmer, G. E., Bolton, J. R., Arnold, D. R.: J. Am. Chem. Soc. *96*, 3708 (1974).

77) Moritani, I., Hosokawa, T., Obata, N.: J. Org. Chem. *34*, 670 (1969).
78) Dürr, H., Schrader, L.: Chem. Ber. *103*, 437 (1970); Angew. Chem. *81*, 426 (1969).
78a) Dürr, H., Schmitz, H.: unpublished results.
78b) Dürr, H., Schmitz, H.: Angew. Chem. *87*, 674 (1975).
78c) Dürr, H., Weiß, B.: Angew. Chem. *87*, 674 (1975).
79) Carstensen-Oeser, E., Müller, B., Dürr, H.: Angew. Chem. *84*, 434 (1972); Angew. Chem. Intern. Ed. Engl. *11*, 422 (1972).
80) Dürr, H., Schmidt, W., Sergio, R.: Liebigs Ann. Chem. *1974*, 1132.
81) Dürr, H., Sergio, R., Gombler, W.: Angew. Chem. *84*, 215 (1972). — V. Brüderle: unpublished.
82) Dürr, H., Ranade, A. C., Halberstadt, I.: Tetrahedron Letters *1974*, 3041. — Ranade, A. C.: (unpublished).
83) Eaton, D. F., Bergman, R. G., Hammond, G. S.: J. Am. Chem. Soc. *94*, 1351 (1972).
84) Gassmann, P. G., Greenlee, W. J.: J. Am. Chem. Soc. *95*, 980 (1973).
85) Berson, J. A., Olin, S. S.: J. Am. Chem. Soc. *92*, 1086 (1970).
86) White, D., Condit, P. B., Bergman, R. G.: J. Am. Chem. Soc. *94*, 7931 (1972).
87) White, D. H., Condit, P. B., Bergman, R. G.: J. Am. Chem. Soc. *94*, 1348 (1972).
88) Condit, P. B., Bergman, R. G.: Chem. Commun. *1971*, 4.
89) Franck-Neumann, M.: Angew. Chem. *80*, 42 (1968); Angew. Chem. Intern. Ed. Engl. *7*, 65 (1968); Tetrahedron Letters *1968*, 2979.
90) Allred, E. L., Smith, R. L.: J. Am. Chem. Soc. *91*, 6766 (1969).
91) Roth, W. R., Ender, K.: Liebigs Ann. Chem. *730*, 82 (1969). Tanida, H., Teratake, S., Hata, Y., Watanabe, M.: Tetrahedron Letters *1969*, 5341, 5345. Tanida, H., Teratake, S.: Tetrahedron Letters *1970*, 4991. Aratani, T., Nakanisi, Y., Nozaki, H.: Tetrahedron *1970*, 4339. Filipescu, N., De Member, J. R.: Tetrahedron *24*, 5181 (1968). Allred, E. L., Hinshaw, J.: J. Am. Chem. Soc. *90*, 6885 (1968).
Prinzbach, H., Stusche, D.: Angew. Chem. *82*, 836 (1970). — Masamune, S., Zenda, H., Wiesel, M., Nakatsuka, N., Bigam, G.: J. Am. Chem. Soc. *90*, 2727 (1968).
92) Pagni, R. M., Watson, C. R., Jr., Bloor, J. E., Dodd, J. R.: J. Am. Chem. Soc. *96*, 2291, 4064 (1974).
93) Sasaki, T., Eguchi, S., Hibi, F.: Chem. Commun. *1974*, 227.
94) Kopecky, K. R., Evani, S.: Can. J. Chem. *47*, 4041 (1965).
95) Pacifici, J. G., Irik, G.: Tetrahedron Letters *1969*, 1331, 2207. Pacifici, J. G., Irik, G., Anderson, C. G.: J. Am. Chem. Soc. *91*, 5654 (1969); see also: Wan, J. K. S., Hess, L. D., Pitts, J. N.: J. Am. Chem. Soc. *80*, 2069 (1964).
96) Badger, G. M., Joshua, C. P., Lewis, G. E.: Australian J. Chem. *18*, 1639 (1965). Badger, G. M., Drewer, R. J., Lewis, G. E.: Australian J. Chem. *19*, 643 (1966). Applequist, D. E., Lintner, M. A., Searle, R.: J. Org. Chem. *33*, 254 (1968).

Received November 12, 1974

Triplet States from Azides

Professor Dr. Heinz Dürr and Dr. Helge Kober

Fachbereich 14 der Universität des Saarlandes, Fachrichtung 14.1
Organische Chemie, D-6600 Saarbrücken

Contents

1. Introduction

Loss of nitrogen from organic azides results in most cases in uncharged, electron-deficient molecular fragments *1*, containing a nitrogen atom with only six electrons in its valence shell.

$$R{-}N_3 \longrightarrow R{-}\underline{\overline{N}} + N_2$$

1

These reactive intermediates were proposed first for the Lossen rearrangement[a] in 1891 by Tiemann [1] and for the photochemical decomposition of hydrazoic acid in 1928 by Beckman and Dickinson [2]. The chemistry in this field, which attracted little interest for a long time, was stimulated by the development of carbene chemistry. In the last fifteen years, beginning with a summary by Kirmse [3] in 1959, excellent reviews on nitrene chemistry appeared [4–12].

No agreement on the nomenclature for these neutral, univalent nitrogen intermediates was achieved for a long time. The names used include imene [14], imine radical [15,16], azene [17], azacarbene [18], imin [19], and nitrene [20]. The nitrene nomenclature, which parallels that for carbenes, is commonly used today and will be employed in this article. In this nomenclature $H{-}\overline{N}{:}$ is nitrene, $C_6H_5{-}\overline{N}{:}$ is phenylnitrene, $CH_3{-}O{-}C(=O){-}\overline{N}{:}$ is methoxycarbonylnitrene etc.

For nitrenes both a singlet state *1a*, with two pairs of electrons and a low-lying empty p_x-orbital, and a triplet state *1b*, with one electron pair and two electrons with parallel spins, are possible.

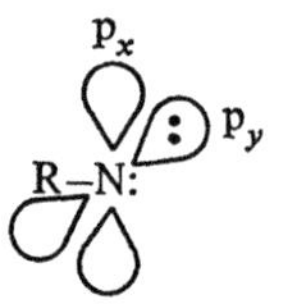

1a ↑↓ (S_0)

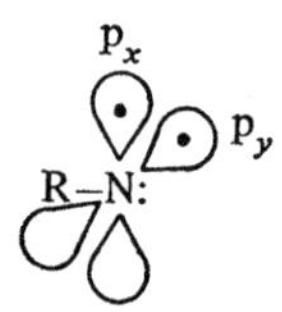

1b ↑↑ (T_1)

With respect to the electronic structure, nitrenes are analogous (isoelectronic) to carbenes. As in the case of diazocompounds, photolysis as well as thermolysis of azides leads to the corresponding nitrenes. Some reactions are collected in the following scheme:

$$HN_3 \xrightarrow{h\nu/\Delta} H\overline{N}{:} + N_2$$

$$RN_3 \xrightarrow{h\nu/\Delta} R{-}\overline{N}{:} + N_2$$

$$R{-}SO_2N_3 \xrightarrow{h\nu/\Delta} R{-}SO_2\overline{N}{:} + N_2$$

---

[a] It now appears unlikely that nitrenes are involved in this rearrangement. Today for this reaction a concerted mechanism for the N_2 expulsion and the migration of the substituent is formulated [11].

$$R{-}CON_3 \xrightarrow{h\nu/\Delta} R{-}CO{-}\overline{N}{:} + N_2$$

$$ROCON_3 \xrightarrow{h\nu/\Delta} ROCO\overline{N}{:} + N_2$$

$$NCN_3 \xrightarrow{h\nu/\Delta} NC\overline{N}{:} + N_2$$

Other methods[b)] (see the following scheme), including

a) deoxygenation of nitro- and nitroso-compounds with triethyl- or triphenyl-phosphite,

b) base-induced α-elimination,

c) oxydation of hydrazine derivatives with mercury oxide or lead tetraacetate and

d) photolysis of oxaziridines

all lead to nitrenes. But the intermediates generated by these procedures will not be included in this article.

$$Ar{-}NO_2 + R_3P \longrightarrow Ar{-}\overset{O^{\ominus}\uparrow}{N}{-}O{-}\overset{\oplus}{P}R_3 \longrightarrow ArNO + R_3PO$$

$$Ar{-}NO + R_3P \longrightarrow Ar{-}\overset{\ominus}{N}{-}O\overset{\oplus}{P}R_3 \longrightarrow Ar{-}\overline{N}{:} + R_3PO \qquad \text{(a)}$$

$$R{-}N\begin{matrix}H\\Ph\end{matrix} + B \longrightarrow R{-}\overline{N}{:} + BH^+ + Ph^- \qquad \text{(b)}$$

$$R_2N{-}NH_2 \xrightarrow{HgO} R_2{-}N{-}\overline{N}{:} \qquad \text{(c)}$$

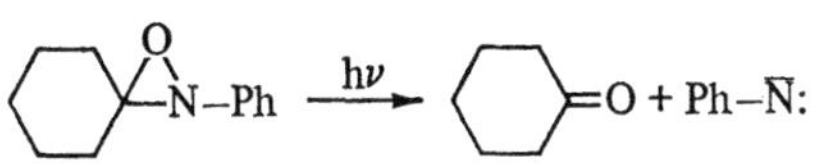

(d)

1.1. Thermolyses

The thermal decomposition of most organic azides will lead to nitrene intermediates which, according to the spin conservation rule, must be formed initially in the singlet state (S_0).

$$R{-}N_3 \xrightarrow{\Delta} \underset{1(S_0)}{R{-}N{:}} + N_2$$

The thermolysis of methylazide was studied first by Ramsperger [21a)] who found it to be a homogenous first-order reaction. Studies of thermolyses of several other organic azides in various solvents and at different concentrations also show first-order kinetics [22–28)]. This led to the conclusion that free nitrene intermediates are involved. There are however thermal decompositions of azides where nitrene intermediates are not involved. For example ortho-substituted arylazides like o-nitrophenylazide *2*, undergo formation of benzofurazan *3* by a concerted mechanism according to l.c.[21b)].

b) For a detailed review see Ref. [9b)].

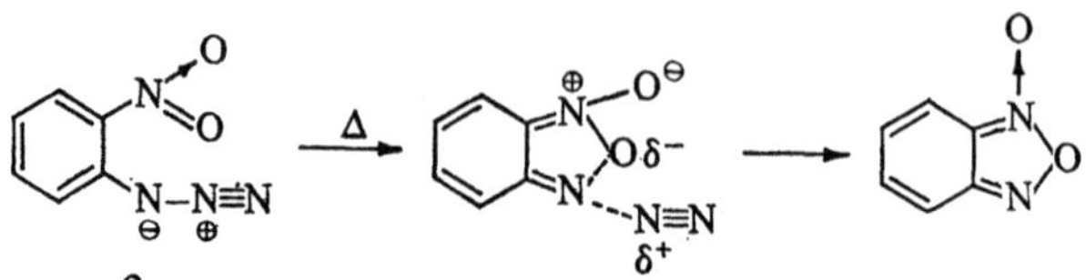

2 3

The thermolysis of arylazides *4* leading *via* a Curtius-rearrangement to isocyanates *5* most probably does not involve a nitrene intermediate.

$$\underset{4}{R{-}CON_3} \xrightarrow{\Delta} \underset{5}{R{-}N{=}C{=}O} + N_2$$

So nitrenes could not be trapped in this reaction by insertion into C—H bonds or olefinic double bonds. Decomposition temperatures of several azides and their corresponding rate constants (k) are listed in Table 1. Unlike most organic azides,

Table 1. Decomposition temperatures and rate constants of various nitrenes

Azide	Decompoposition Temperature	Solvent	$k \times 10^4$ [sec^{-1}]	Ref.
C_6H_5 $C_6H_5{-}C{-}N_3$ C_6H_5	180	Dibutylcarbitol	2.4	[22]
C_6H_5 $CH_3{-}C_6H_4{-}C{-}N_3$ C_6H_5	180	Dibutylcarbitol	2.7	[22]
C_6H_5 $O_2N{-}C_6H_4{-}C{-}N_3$ C_6H_5	180	Dibutylcarbitol	2.8	[22]
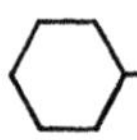$-N_3$	194	Ethylbenzoate	0.38	[23]
$C_6H_5{-}N_3$	132	Ethylbenzoate	0.078	[23]
p—$CH_3O{-}C_6H_4{-}N_3$	132	Ethylbenzoate	0.91	[23]
m—$CH_3O{-}C_6H_4{-}N_3$	174	Aniline	8.8	[24]
m—$CH_3{-}C_6H_4{-}N_3$	174	Aniline	8.8	[24]
$CH_3{-}C_6H_4{-}SO_2{-}N_3$	155	Diphenylether	3.4	[25]
$H_3C{-}(CH_2)_3{-}CH_2{-}SO_2{-}N_3$	166	Diphenylether	4.4	[25]
$H_3C{-}CH_2{-}CH_2{-}O{-}CO{-}N_3$	130	Diphenylether	5.9	[26]
$H_3C{-}(CH_2)_{16}{-}CH_2{-}O{-}CO{-}N_3$	120	Diphenylether	2.4	[27]
$N{\equiv}C{-}N_3$	40—50	n-Hexane	~0.3[1]	[28][1]

[1] Estimated value from the reaction time given in Ref. [28].

which require temperatures for the fragmentation to the corresponding nitrenes above 100°, cyanogenazide smoothly evolves nitrogen when heated to 40–50 °C. This fact is due to the symmetry and the resonance stabilisation of the resulting cyanonitrene.

Thermolysis, as pointed out earlier, produces nitrenes being initially in their singlet state. Triplet reactions will be observed therefore only when time for spin inversion (isc) is available before the nitrene reacts.

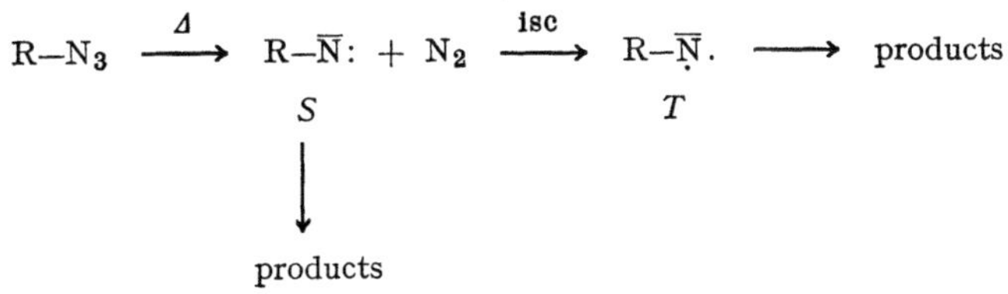

1.2. Photolyses

Photolysis of organic azides can lead to the nitrene in the singlet or triplet state.

$$\mathrm{R{-}N_3} \xrightarrow[-\mathrm{N_2}]{h\nu} \underset{S}{\mathrm{R{-}\overline{N}:}} \quad \text{or} \quad \underset{T}{\mathrm{R{-}\overline{\dot{N}}.}}$$

The wavelength required for effective absorption of UV light by azides depends on the substituents. The azides absorb normally below 290 nm, with the exception of arylazides, which show also transitions between 300 and 350 nm. Therefore as light sources low-pressure mercury-lamps are most commonly used. In Table 2 the range of absorption, the ε-values and the quantum yields of some azide photolyses are compiled.

Photolysis of an azide will lead to an excited singlet state of the azide which can:

1. undergo intersystem crossing to a triplet azide,
2. go back to starting material by internal conversion in a radiationless decay process,
3. decompose to a singlet nitrene.

In addition, the singlet nitrene can undergo intersystem crossing to the triplet nitrene. The T_1-azide state can thus be populated by two routes (a) and (b). These possibilities are indicated in the following scheme:

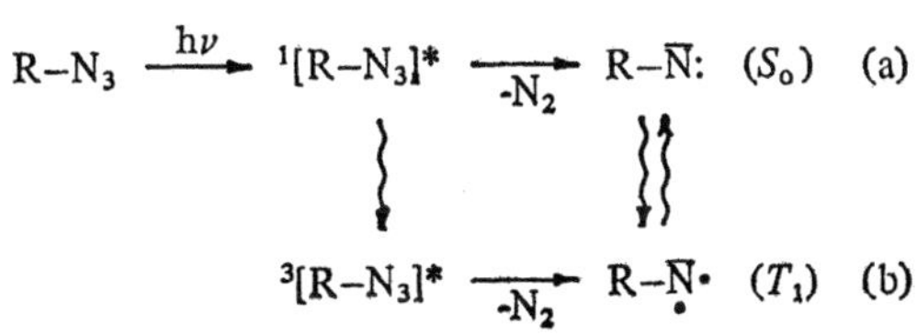

Table 2. UV-spectra of azides

Compound	λ_{max} (log ε)	λ of light used	Quantum yield	Ref.
CH_3—CH_2N_3	—	253 313	0.88[4] 0.89[5]	31,32)
CH_3—CH_2—CH_2—N_3	—	313	0.83[4]	31,32)
CH_3—$(CH_2)_2$—CH_2—N_3	216(2.6) n—π^* 287(1.3) π—π^*	313	0.78[1]	31–33)
CH_3—$(CH_2)_4$—CH_2—N_3	—	253 313	0.86[4] 0.71[4]	31,32)
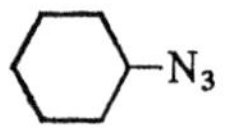	216(2.6) n—π^* 287(1.3) π—π^*	313	0.68	31–33)
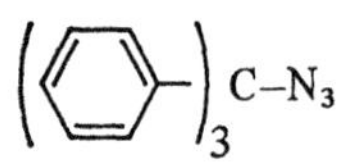	—	313	0.80[4]	31,32)
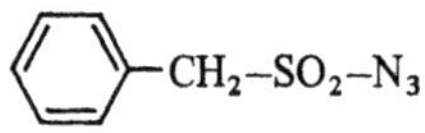	217(4.7)	—	—	36)
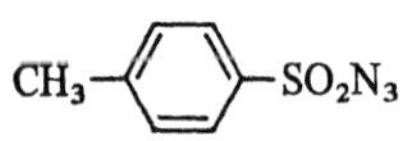	232(4.3)	—	—	36)
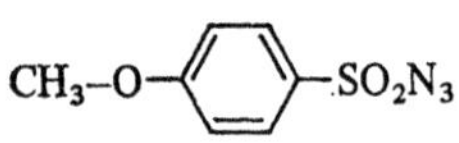	248(4.9)	—	—	36)
CH_3—SO_2—N_3	254(1.1)	—	—	37)
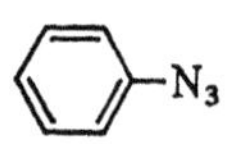	247(3.9) π—π^* 278(3.3) π—π^* 315(1.9) n—π^*	254	0.53[1] 0.52[2]	29,30)
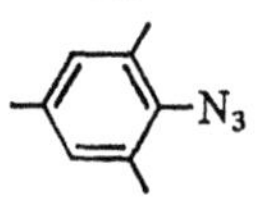	247(3.4) π—π^* 286(2.8) π—π^* 300(2.0) n—π^*	254	0.96[1] 0.98[2]	29,30)
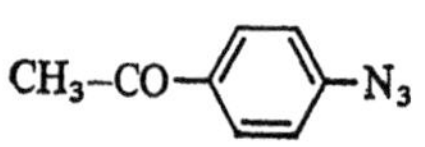	—	280	1.00[3] 1.00[2]	29)
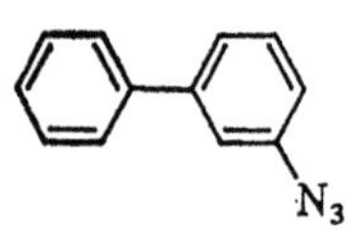	246(4.2) π—π^* 292(3.5) π—π^* 320(2.3) n—π^*	254	0.37[1] 0.36[2]	29,30)
CH_3CH_2—O—C(=O)—N_3	254(1.9)	—	—	35)
NCN_3	220(3.3) 275(2.2)	—	—	34)

1) Hexane solution.
2) Methylcyclohexane-Isopentane matrix.
3) Cyclohexane solution.
4) Methanol solution.
5) Ethylether solution.

An exception is the unsensitized photolysis of tertiary alkylazide in which according to Abramovitch and Kyba [13] a discrete nitrene intermediate is not formed.

Depending on the substituent R and the reaction medium, the photochemically generated nitrene can undergo different reactions. These will be analogous to those observed in the thermolysis of azides but may differ in product ratios.

2. Electronic States and Structure of Triplet Nitrenes

2.1. Calculations

The electronic structure of the parent nitrene NH, which is also the best known and understood, has been calculated by MO and valence-bond methods [38]. Various molecular orbital methods such as LCAO-SCF method [39,40], SCF-MO-calculation allowing for electron correlations [41] and configuration interaction [42,43] have been used to calculate electronic structure of ground and excited states of nitrenes. The molecular ground state is a $^3\Sigma^-$triplet (like CH_2) and the lowest excited state is the $^1\Delta$ state, being 1.76 eV above the ground state. Both states have the same molecular orbital configuration $1\sigma^2\ 2\sigma^2\ 3\sigma^2\ 1\pi^2$. In the ground state, the two degenerated orbitals p_x and p_y are occupied by one electron each, thus having a total spin number of $S=1$. The multiplicity is therefore $2\,S+1=3$ ($^3\Sigma$-triplet state). If the two orbitals are not degenerated then two electrons are present in the same orbital and the multiplicity now equals 1 ($^1\Delta$ singlet state). Calculations of the electronic states and electronic spectra of the substituted alkylnitrenes R$\overline{\text{N}}$| are rather rare. A theoretical study of the electronic structure of methylnitrene has recently been published [44] and the lowest electron configuration was found to be $1a_1$, $^22a_1$, $^23a_1$, $^34a_1$, ^21e, $^45a_1$, $^22e^2$. Besides |$\overline{\text{N}}$H, extensive calculations have been done only for cyanonitrene [45] and carbonylnitrenes [46].

For cyanonitrene it was shown that the linear structure is favored over the bent one and the ground state is a triplet

·N=C=N·

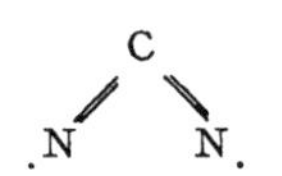

($2\sigma_1$, $2\sigma_2$, $2\sigma_3$, $4\pi_1$, $2\sigma_4$, $2\pi_2$). The electronic configuration of several carbonylnitrenes has recently been determined by nonempirical molecular orbital studies [46]. The calculations suggest that acylnitrenes RCO$\overline{\text{N}}$: are *ground state triplets* while carbalkoxynitrenes ROCO$\overline{\text{N}}$: may have *singlet ground states*. The last point, however, must await further work for clarification.

For arylnitrenes only qualitative treatments have been carried out by Reiser *et al.* [30]. It was concluded that the ground state of arylnitrenes is also a triplet as expected from the experimental results. No quantum mechanical calculations have been reported for sulfonylnitrenes, but ESR-evidence indicates a triplet ground state also for these intermediates [47] (see below).

With the exception of carbalkoxynitrenes the calculations indicate a *triplet ground state* for *all* nitrenes. However the triplet ground state of a nitrene may not be achieved if the nitrene is undergoing a very fast reaction, *i.e.* a nitrene may then react in its singlet state.

2.2. Electron Spin Resonance (ESR)

This technique is the best one to prove the triplet multiplicity of a nitrene. The nitrenes investigated so far were always generated by photolysis of the corresponding azides at low temperatures (4 or 77 K) in randomly oriented glasses. The resulting ESR-spectra show transitions between spin states ($S = 1, 0, -1$) of a reactive species with two unpaired spins. The spectrum can then be described by the following equation:

$$\mathcal{H} = g \cdot \beta \cdot HS + DS_z^2 + E(S_x^2 - S_y^2)$$

where: S = spin operator, g = Landé splitting factor and D and E are the so-called zero field splitting parameters, which should be present in the absence of an external field. The value of D is roughly proportional to the separation of the two unpaired spins, thus being a measure for electron delocalization. E indicates the deviation of the spin interaction from cylindrical symmetry.

In the absence of external perturbation, photodecomposition of azides must occur with overall conservation of spin. Within this restriction there are four possibilities for the fragmentation:

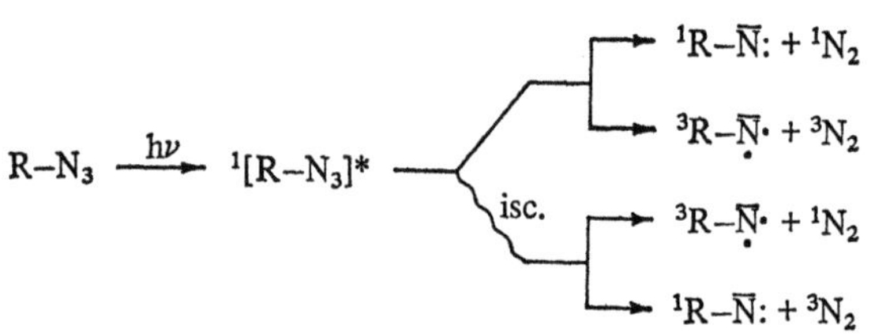

(1)

(2)

(3)

(4)

Routes (2) and (4) are unlikely since the light employed in the photolysis does not have enough energy to generate molecular nitrogen in the triplet state. An equilibrium, however, between the singlet and triplet nitrene formed according to route (1) and (3) is theoretically possible (see above). If there is enough time, the reactive species initially formed can equilibrate with its ground state provided that it does not react before intersystem crossing (isc) could occur.

$$\underset{\textit{1a}\ (S_0)}{R\text{–}\bar{N}:} \rightleftarrows \underset{\textit{1b}\ (T_1)}{R\text{–}\dot{\bar{N}}\cdot}$$

The theoretical calculations mentioned above predict a T_1-ground state *1b* for nitrenes and this is in agreement with many ESR-measurements, where the triplet species can be observed. The ESR-spectra of nitrenes generated at 77 K in a matrix by UV-irradiation of the corresponding azides were detected and assigned to the triplet ground state for the following species: C_6H_5–$\dot{\bar{N}}\cdot$; o—CF_3—

Table 3. Zero field splitting parameters for nitrenes

Nitrene	D (cm^{-1})	E (cm^{-1})	Ref.
$CH_3{-}\bar{N}\cdot$	1.595	0.00	48)
$CH_3{-}CH_2{-}CH_2{-}\bar{N}\cdot$	1.607	0.0034	44)
$CH_3{-}(CH_2)_5{-}CH(CH_3){-}\bar{N}\cdot$	1.616	0.0019	44)
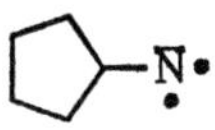	1.575	<0.002	44)
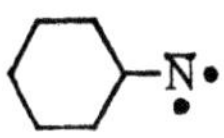	1.599	<0.002	44)
$(CH_3)_3C{-}\bar{N}\cdot$	1.625	<0.002	44)
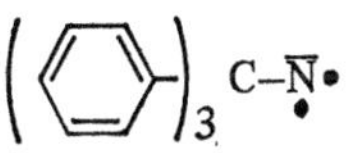	1.660	0.00	48)
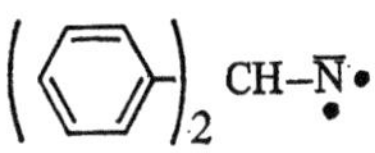	1.636	0.00	48)
$CH_3{-}SO_2{-}\bar{N}\cdot$	1.581	0.0036	44)
$NC\bar{N}\cdot$	1.544	<0.002	49)
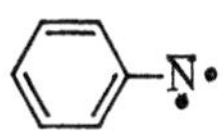	0.99	<0.002	44)
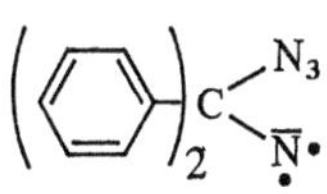	1.625	0.00	48)
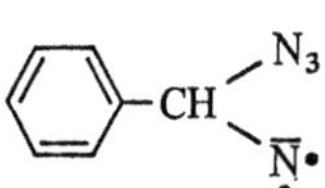	1.626	0.00	48)
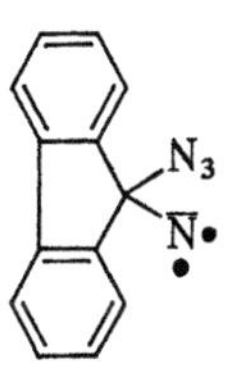	1.606	0.003	50)

C_6H_4—$\dot{\overline{N}}$·; C_6H_5—SO_2—$\dot{\overline{N}}$·; p—CH_3—C_6H_4—SO_2—$\dot{\overline{N}}$· and NC$\dot{\overline{N}}$· [47,49]. Attempts to observe triplet alkylnitrenes by direct irradiation of alkylazides at 77 K were unsuccessful. At lower temperature (4 K) however, weak signals were observed [48]. So Wasserman, Smolinski and Yager [48b] could show by ESR-spectroscopy that in analogy to H$\dot{\overline{N}}$· the ground state of alkylnitrenes is a triplet. No ESR-signals were observed for C_6H_5—CH=CH—$\overline{N}$|, CH_3—CH_2—O—CO—$\overline{N}$| and C_6H_5—O—CO—$\overline{N}$|; probably they react too fast to be present in a detectable stationary concentration [47,49]. On the other hand the ground state may be a singlet as discussed in the case of R—O—CO—$\overline{N}$: [46]. In Table 3 the D- and E-values of some nitrenes are collected.

The smaller the value of D, the smaller the interaction between the unpaired electrons, or in other words the greater is their separation. So, from the zero field splitting parameter of phenylnitrene ($D = 0.99$ cm^{-1}) one can deduce that one unpaired electron is delocalized into the aromatic system. Although ESR shows that these nitrenes have triplet ground states it tells nothing about the electronic state of the nitrenes at the moment of generation.

2.4. Electronic Spectra

The UV and IR spectra for the parent :$\overline{N}$H have been measured by the flash photolysis technique as well as with the matrix technique at low temperature (for detailed reviews see Abramovitch and Davies [5] and Berry [38]). For :$\overline{N}$H four spectral bands are detected. The sharp $A\,^3\pi \leftrightarrow X^3\Sigma^-$ transition is readily observed at approximately 336 nm in the absorption and emission spectra. The broader $c^1\pi \rightarrow a^1\Delta$ transition is well known in emission, but difficult to observe in absorption. Emission from the $c^1\pi \rightarrow b^1\Sigma^+$ and $d^1\Sigma^+ \rightarrow c^1\pi$-transitions appears as a 453 nm and 254 nm band respectively.

For the homologous alkylnitrenes no electronic spectra have been reported so far.

The optical properties of arylnitrenes are much better known than for alkylnitrenes; this is largely due to the optical spectroscopy of azide photolysis products by Reiser *et al.* [30,51–54]. The arylnitrene UV-spectra reported there show three main absorption bands. Typically these have maxima in the 370—390 nm, 310—370 nm and 220—260 nm regions. The triplet arylnitrenes give rise to spectra resembling those of aromatic radicals. Thus phenylnitrene in an organic glass at 77 K exhibits two bands at 314 and 402 nm which correspond to the bands of the benzyl radical at 318 and 463 nm. The UV- and IR-spectra of NC$\overline{N}$: have been reviewed by Anastassiou *et al.* [45]. Flash photolysis of NCN_3 produces a high concentration of the nitrene which shows the characteristic absorption of this species in the 329 nm region. A much further information on NC$\overline{N}$: was obtained by Milligan *et al.* [55,56] on the basis of matrix isolation studies. They generated NC$\overline{N}$: by photolysis ($\lambda > 280$ nm) of NCN_3 and observed a characteristic absorption at 329 nm, which they assigned to a $^3\Sigma_g^- \rightarrow {}^3\Sigma_u^-$-transition. In addition to this band, a series of bands appears between 300 and 240 nm. In the infrared NC$\overline{N}$: displays a band at 423 cm^{-1} as well as an intense absorption at 1475 cm^{-1}. Energy considerations suggest that the 423 cm^{-1} band is due to a bending vibration (ν_2) and the 1475 cm^{-1} absorption to an antisymmetric stretching vibration (ν_3).

3. Reactions of Triplet Nitrenes

The typical reactions of nitrenes generated from aliphatic azide precursors are shown below and include isomerization to imines (a), intramolecular insertion into C—H bonds (b) has been questioned (see l. c. [11] p. 57) and intermolecular hydrogen abstraction (c).

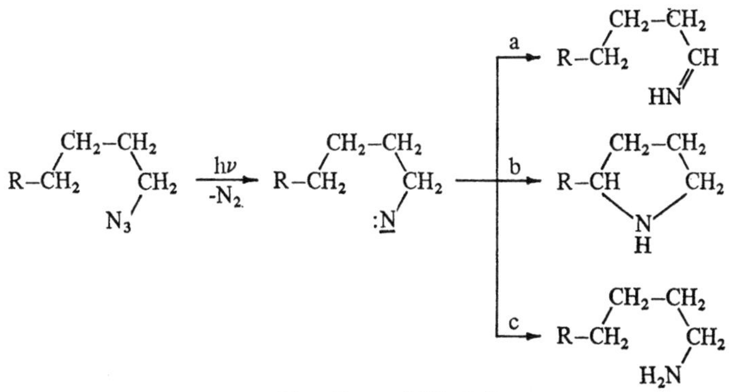

Reactions of Alkylnitrenes

General reactions of nitrenes — where R = ROC—, R—SO_2—, Ar— and NC— are insertion into sigma bonds (C—H, N—H, O—H) (a), addition to double bonds (b), dimerizations (c) and H-abstractions (d).

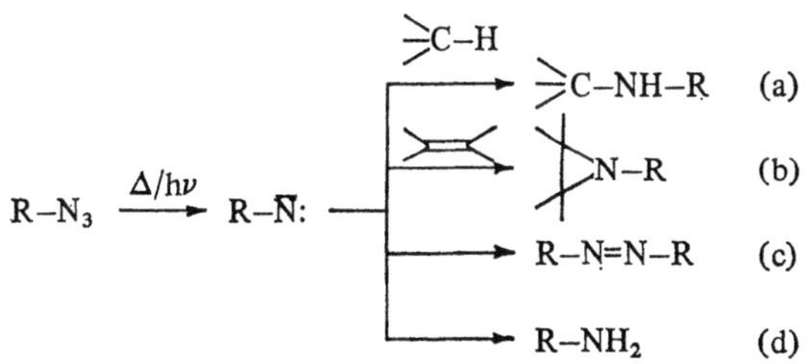

Reactions of Nitrenes (R = see above)

These modes of stabilization are not available for all nitrenes. The reactions observed depend on the nature of the R-group, on the reaction conditions, as well as on the spin state of the nitrene. In this article only reactions which involve a triplet nitrene will be discussed.

3.1. Abstraction Reactions

Nitrenes can formally insert in a number of σ-bonds. This is a common reaction which in some cases will be the dominant if not the exclusive one. Three different mechanisms can be postulated for this reaction as pointed out in the following scheme.

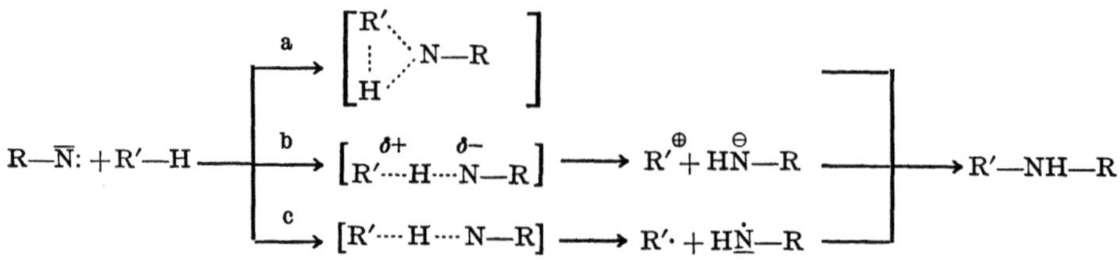

Of these mechanistic possibilities (a) represents the commonly considered one-step reaction of a *singlet nitrene*. Path (c) is a two-step reaction indicating a *triplet nitrene*. Mechanism (b), involving the *ylid formation* is possible with highly electrophilic species such as NC$\overline{\text{N}}$: only, which possess a vacant low-lying bonding MO, *e.g.* π_g of S_2 [45]. *Hydrogen abstraction* in aliphatic solvents probably involves the *triplet nitrene* [25,57] which reacts with loss of configuration, while singlet nitrenes are thought to insert stereospecifically into C—H bonds [6,25]. The abstraction reactions of nitrenes are therefore an indication of a triplet species. Unfortunately alkylnitrenes undergo 1,2-H-migration with great facility (if there is no H attached to α-carbon then 1,2-alkyl- or 1,2-aryl-migration will occur) to the corresponding imines. Therefore other "nitrene products" are formed only in small amounts. At the most, a few percent amine (from intermolecular abstraction) is observed, even in the presence of hydrogen donors. Rapid gas phase pyrolysis (in argone) of some alkylazides [58] gave the corresponding amines.

$$R\text{—}N_3 \xrightarrow{\Delta} R\text{—}NH_2 + \text{other products}$$

Addition of hydrocarbons to the reaction mixture does not change the amine yields noticeably. Similiar results are obtained by solution-phase photolysis of alkylazides [59]. The amine yields of these experiments are collected in Table 4.

Table 4. Gas phase pyrolysis and solution photolysis results of various alkylazides

Azide	Gas phase pyrolysis	Gas phase pyrolysis azide + cyclohexane 1:7	Solution phase photolysis	Ref.
CH_3—N_3	1%	1%	—	58)
CH_3—CH_2—N_3	<1%	—	—	60)
CH_3—$(CH_2)_2$—CH_2—N_3	1%	1%	5%	58,59)
CH_3—$(CH_2)_6$—CH_2—N_3	1%	—	12%	58,59)
(CH_3—CH_2)(CH_3)CH—N_3	3%	—	—	58)
CH_3—CH(CH_3)—CH_2—CH_2—CH_2—N_3	1%	—	—	58)
CH_3—CH(Ph)—CH_2—CH_2—N_3	—	—	7%	58,59)

Abstraction of hydrogen by the triplet nitrene to form primary amines is perhaps the most general reaction of arylnitrenes in solution. The source of hydrogen may be the solvent or the nitrene precursor. Although many arylazide thermolyses were carried out under comparable conditions (see Table 5), there is no study available which deals with the relation between nitrene structure and the reaction paths which are possible. From the experimental data it can also be seen that the extent of hydrocarbon abstraction depends on the reaction medium, but no systematic study has as yet been reported.

Table 5. Thermolysis of arylazides to triplet-abstraction products

Azide	Solvent	Temperature [°C]	Ar—NH_2(%)	Ref.
C_6H_5—N_3	p-Xylene	138	20—25	[61]
C_6H_5—N_3	Benzene	155	20—25	[61]
C_6H_5—N_3	n-Pentane	160—180	30	[63]
C_6H_5—N_3	i-Pentane	160—180	15	[63]
C_6H_5—N_3	Thiophenol	165	52	[64]
C_6H_5—N_3	Decalin	141	44	[62]
m—Br—C_6H_4—N_3	Decalin	141	85	[62]
m—CH_3—O—C_6H_4—N_3	Decalin	141	97	[62]
m—O_2N—C_6H_4—N_3	Decalin	141	1	[62]
p—O_2N—C_6H_4—N_3	Decalin	141	< 1	[62]
p—Br—C_6H_4—N_3	Decalin	141	78	[62]

The nature of hydrogen abstraction process is normally indicated by the fact that the hydrogen-supplying solvent undergoes dehydrogenative coupling to a considerable extent but always in a poorer yield as the aniline formation would predict. Thermolysis of phenylazide *6* in p-xylene *7* affords aniline (*8a*) and syn-di-p-tolylethane (*9*) [61]. Thermolysis of p-methoxyphenylazide (*10*) in cumene (*11*) gave the aniline *12* and bicumyl (*13*) [23].

Smith [65] concluded the strong preference of hydrogen abstraction by the nitrene from the starting material rather than from alkane solvents from the results of the thermolysis of $C_6D_5N_3$ in pentane [66]. The two main products were aniline, which was found to contain 18% of N-D-aniline.

Thermolysis of $C_6H_5N_3$ in $(CH_3)_3CD$ produced aniline containing only 7% N—D, in spite of the fact that the tertiary position of isobutane is the most readily attacked one. Intermolecular insertion of arylnitrene in C—H bonds has been reported recently [63] in the thermolysis of phenylazide in alkane in yields up to 10%, along with aniline as the major product. Hall and co-workers [66] have shown that thermolysis of C_6H_5—N_3 in optical active 2-phenylbutane led to 2-anilino-2-phenylbutane with 40% retention of optical activity; they obtained further information on the nature of the insertion and hydrogen abstraction process from the thermolysis of ArN_3 in mixed solvents. Aniline and N-cyclohexylaniline (ratio

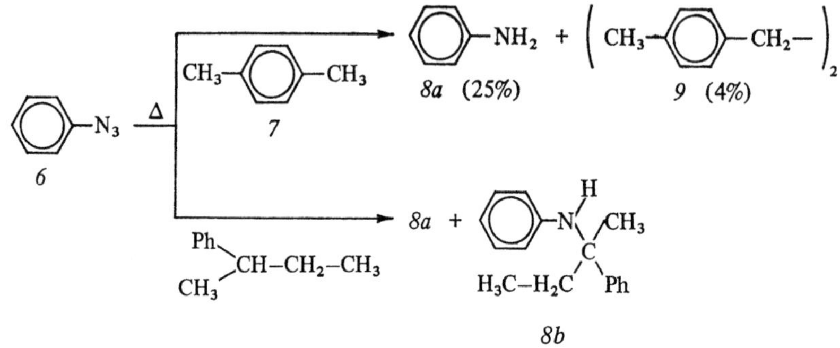

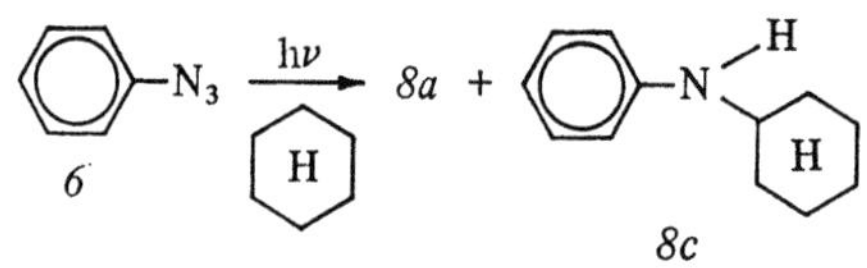

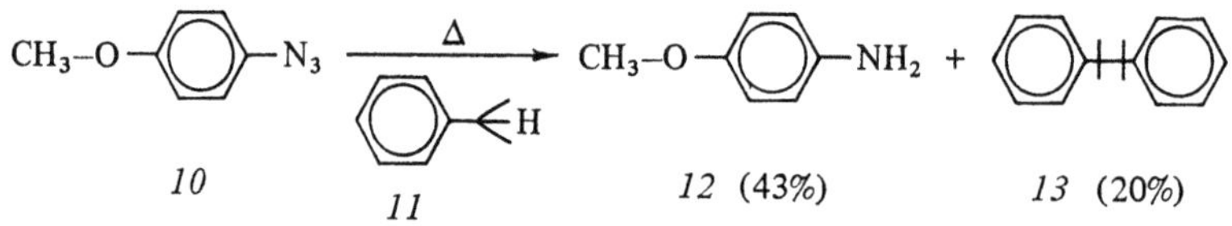

about 4:1) are formed in the photolysis of phenylazide in cyclohexane and it was shown that the product ratio is independent over a wide range of the azide concentration. Similiar results were obtained with p-tolylazide and with p-chlorophenylazide. The inert solvent however should increase the ratio of triplet versus the singlet nitrene available for reaction with cyclohexane. The insensitivity of the product ratio *8a*:*8c* to the increase of inert diluent shows that both products derive from the same electronic state of the nitrene. These results are most easily explained by the following scheme:

$$\text{Ar–N}_3 \xrightarrow{h\nu} \underset{S_1}{\text{Ar–}\overline{\text{N}}:} \xrightarrow{\text{i.s.c.}} \underset{T_1}{\text{Ar–}\overline{\text{N}}\cdot} \xrightarrow{\text{RH}} \underset{\text{solvent cage}}{[\text{Ar–}\overline{\text{N}}\text{H} + \cdot\text{R}]} \xrightarrow{[\text{H}]} \text{Ar–NH}_2$$

$$\downarrow$$

$$\text{Ar–NH–R}$$

The initial step for both products is an abstraction of hydrogen by a triplet nitrene from the hydrocarbon to form an aminoradical and an alkylradical. These radicals originate in the solvent cage, and if they recombine before leaving it, N-alkylaniline (ArNHR) is produced. Escape of the two radicals from the solvent cage after further abstraction from the solvent or the nitrene precursor leads to the amine ($ArNH_2$) as well as to coupling (R—R) and other products. This shows that in the case of arylnitrenes not only the arylamines arise *via* the triplet nitrene but also the alkyl-anilines, which appear to be formed by an insertion process.

Alkylsulfonylnitrenes, generated in aliphatic hydrocarbons undergo hydrogen abstraction to a small extent only. In the thermolysis of 2-propanesulphonylazide

(*14*) in cyclohexane 3% of 2-propanesulphonamide (*15*) were found [25], whereas no abstraction product was detected in the case of 1-pentane-sulfonylnitrene. Thermolysis of α-toluenesulphonylazide (*16*) in n-dodecane [67] afforded 25% of α-toluenesulphonamide (*17*), and 1-pentylsulphonylnitrene gave a 1% yield of dicumyl by H-abstraction from the α-C-atom of cumene [25]. Methanesulphonylazide (*18*), when thermolyzed in toluene, led to 22.7% of methanesulphonylamide and 0.5% of dibenzyl (*13*) [68].

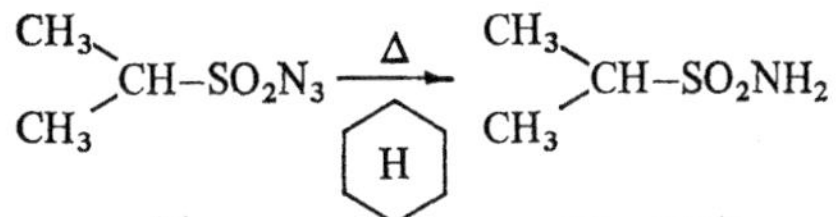

14 *15* (3%)

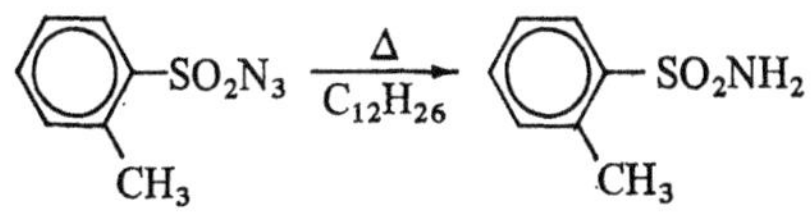

16 *17* (25%)

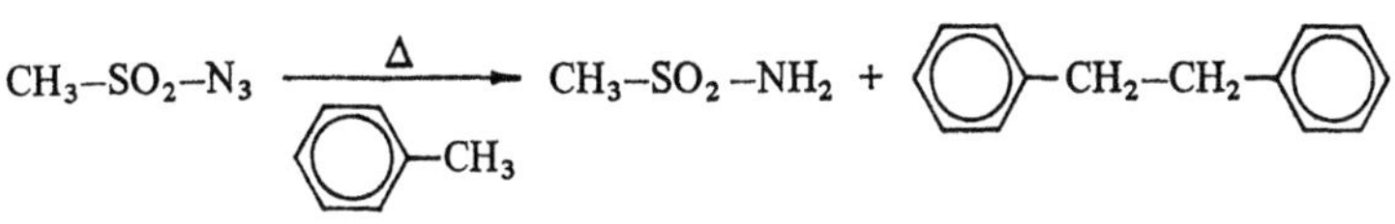

18 *19*(22%) *13*(0,5%)

If arylsulphonylnitrenes are generated in aliphatic hydrocarbons, better yields of hydrogen abstraction products as in the case of alkylnitrenes are obtained. Thermolysis of p-toluenesulphonylazide in cyclohexane [25] gave 5% yield of p-toluenesulphonylamide, whereas 40% yield was reported also [58]. Thermal decomposition of o-substituted phenylazides *20, 21, 22* in dodecane *via* abstraction afforded amides *23, 24, 25.*

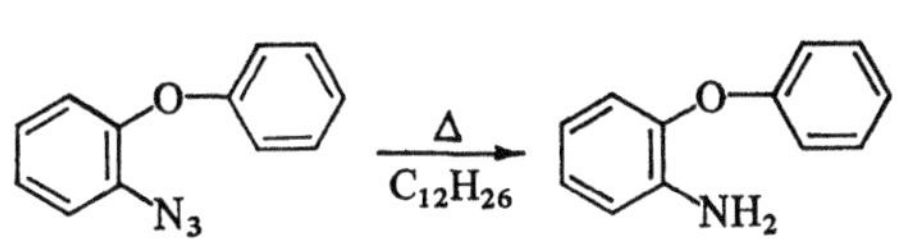

20 *23*(4%)

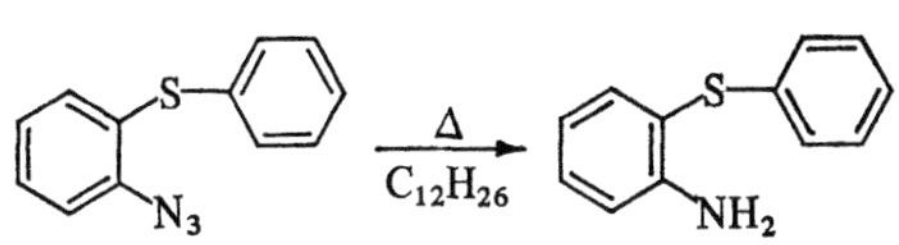

21 *24*(19%)

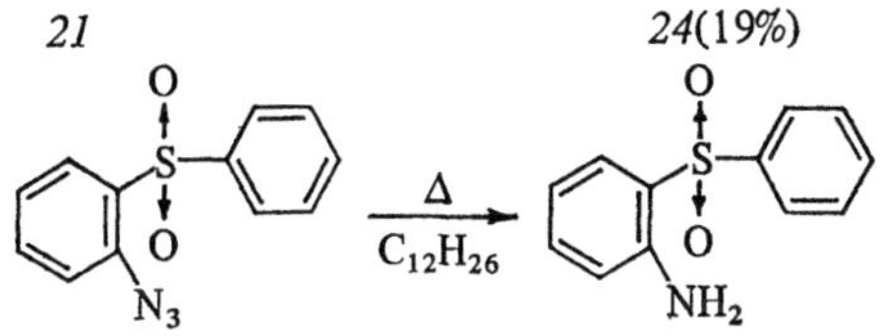

22 *25*(9%)

In contrast to the nitrene reactions with aliphatic hydrocarbons, where only little amide formation occurs [25,69], in aromatic hydrocarbons much better yields of sulphonylamides are obtained [16,68,70]. For example, thermolysis of benzenesulphonylazide and methanesulphonylazide in benzene gave 18% of benzenesulphonamide [16] and 14% ethanesulphonylamide, respectively [68]. In contrast to the nitrene reactions in aliphatic hydrocarbons no coupling product such as biphenyl-derivatives was detected. This should have been the case however if abstraction of a single hydrogen atom did occur thus giving rise to an aryl radical as shown below.

$$R{-}SO_2\ddot{N}\cdot + C_6H_6 \longrightarrow R{-}SO_2\ddot{N}{-}H + [C_6H_5]\cdot$$

$$[C_6H_5]\cdot + C_6H_6 \rightarrow [(C_6H_5{-}C_6H_6]\cdot \xrightarrow{R{-}SO_2\dot{\ddot{N}}H} C_6H_5C_6H_5 + R{-}SO_2NH_2$$

The first step in the reaction is therefore addition of the nitrene to the aromatic double bond forming an aziridine *26* [71]. This intermediate can now collapse to the sulphonamide *27* and an aryne which produces tars [68]. Ring opening to form an aniline derivative *28* — a formal insertion product — is another possibility of stabilization.

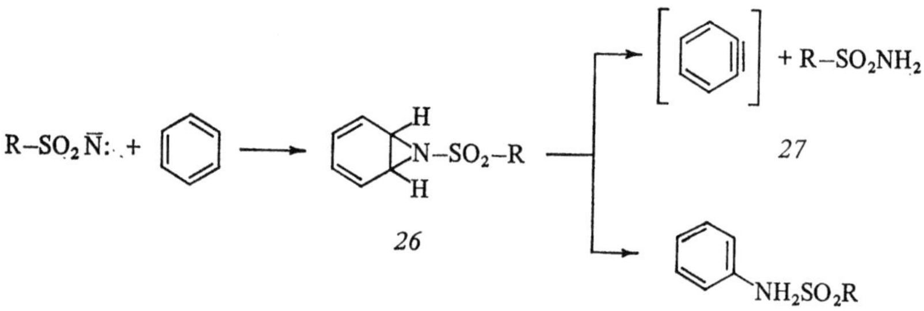

27

26

28

If this mechanism is really operative, the "abstraction product" formed in benzene is not necessarily due to a triplet nitrene precursor. Recently a careful study of the thermolysis of methylazide in substituted benzenes demonstrated that the unsubstituted primary sulfonamide is a product of hydrogen abstraction by the nitrene [72]. On the other hand there are remarkable differences in isomer ratios (o:m:p) of the ring-substituted anilides formed depending on the spin state of the reacting nitrene. The triplet was shown to attach the aromatic nucleus mainly in the o-position, as is expected from a highly electrophilic diradical. Dehydrogenations by carbonylnitrenes have been reported by several authors for a variety of systems. In the direct photolysis of ethylazidoformate *29* in cyclohexene, the amide *30* and the bicyclohexenyl *31* were isolated [35]. Both products result from an abstraction reaction.

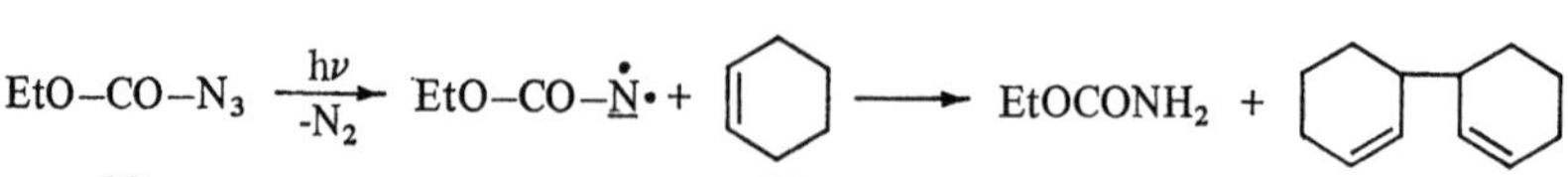

29 *34* *30* *31*

Thermolysis of n-octadecylazidoformate (*32*) in cyclohexane gave 22% amide (*33*) and about 10% cyclohexene *34* [73].

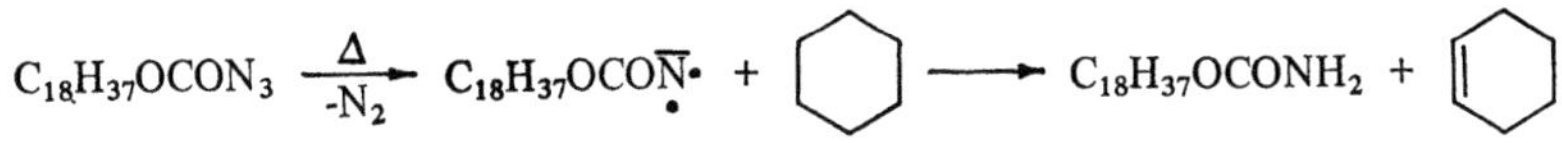

32 *33* *34*

It was suggested that both hydrogens in cyclohexane are removed simultaneously.

3.2. Addition Reactions of Triplet Nitrenes

In contrast to the carbene chemistry, where addition to a double bond is a very common process, in nitrene chemistry the [1+2]-cycloaddition obviously occurs with carbethoxy- and cyanonitrene in an easy way only. There are three main reasons why [1+2]-cycloaddition of nitrenes to carbon-carbon double bonds is relatively rare:

1. The addition products are very unstable at the temperature used for the generation of the nitrenes.

2. The nitrene precursor (usually the azide) undergoes 1.3-dipolar addition to give triazoline, which can loose nitrogen, but does not involve a free nitrene.

3. The olefinic double bond participates in the nitrogen elimination and no free nitrene is formed.

In the case where free nitrenes are involved, both the singlet and the triplet nitrene will add to the double bond. On the basis of Skell's postulate which is now well established for the [1+2]-cycloaddition of carbenes to double bonds [74],

a) *singlet nitrenes* should add *stereospecifically* and

b) *triplet nitrenes* should add *nonstereospecifically.*

This is shown in the following scheme:

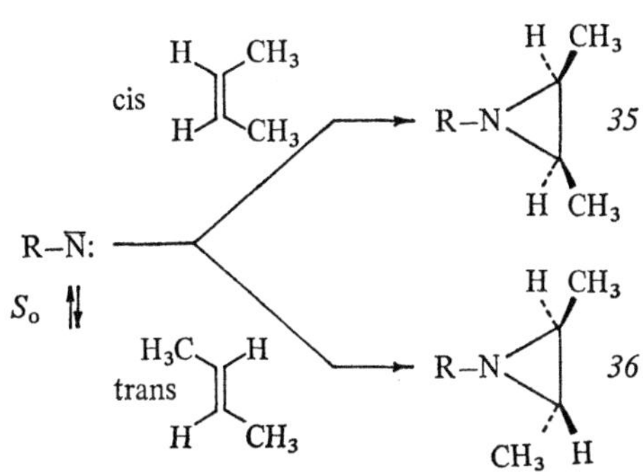

Stereospecific addition

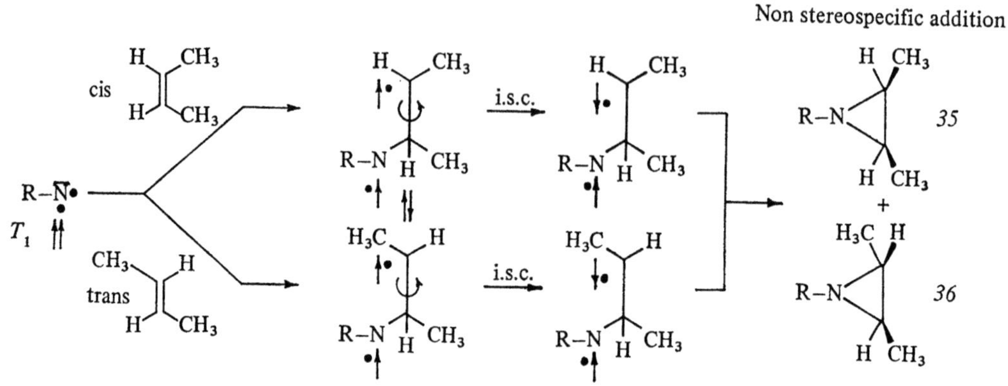

35

36

The addition of a *singlet nitrene* to a double bond is supposed to occur *concertedly*, while the *triplet nitrene* is supposed to add *in two discrete steps via* a 1.3-diradical intermediate. The rate of ring closure of the diradical is supposed to be smaller than that for the rotation around the C—C-bond, therefore the stereochemistry of the olefin will be lost.

Arylazides normally add to olefinic double bonds faster (forming the corresponding triazolines) than their thermal decomposition to nitrenes. These triazolines however loose nitrogen more easily [75–78] (thermally or photolytically) than the corresponding azides and lead to aziridines, which should also arise from a nitrene reaction.

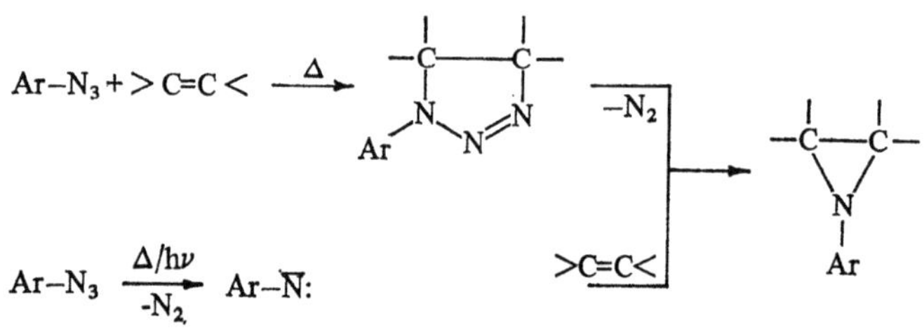

Aziridine formation can therefore never be a proof for the involvement of a nitrene. Addition of carbethoxy nitrenes to carbon-carbon double bonds is observed in solution [35,79–83] as well as in the gas phase [84,85]. In these reactions the singlet and the triplet species are involved, the former add stereospecifically [80,81,83] and the latter nonstereospecifically. The addition of singlet and triplet carbethoxynitrene *29'* to cis- and trans-4-methylpentene-2 *37* has been studied very carefully [80,83,86]. Both adducts *38* and *39* were formed — showing that the reaction was not stereospecific.

From the ratios of cis- and trans-N-carbethoxy-2-methyl-3-isopropylaziridines (*38*) and (*39*) it was concluded that 30% of the photochemically generated nitrene was formed in the *triplet state* [86]. With α-methylstyrene (*40*), which was used as triplet trap, carbethoxynitrene does not lead to aziridines. As 1:1 adduct 3-carbethoxyamino-2-phenyl-propene-1 (*42*) was isolated [83], which was pre-

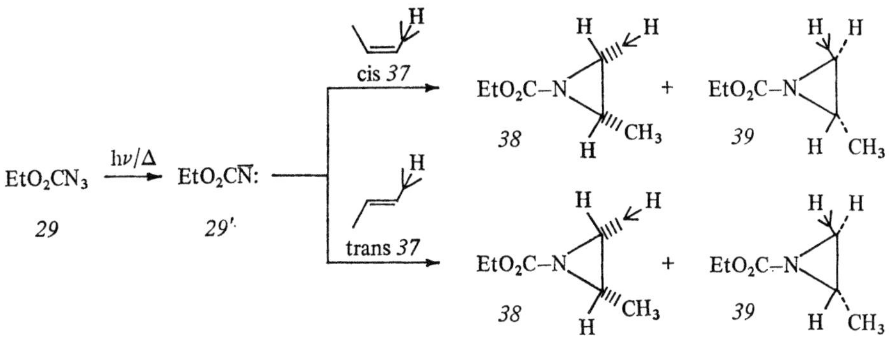

sumably formed by a diradical intermediate *41* and not by direct insertion into the methyl-group.

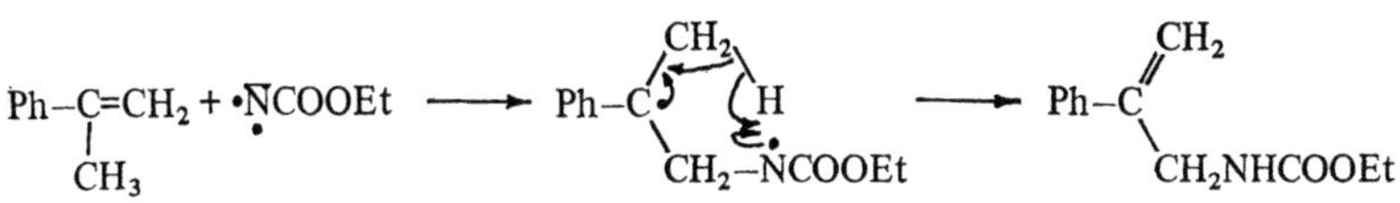

40 41 42

It was however not rigorously excluded that this product was formed from the corresponding aziridine.

With 1,3-dienes only 1,2-addition by nitrenes is observed. In the reaction of carbethoxynitrene with isoprene *43* in contrast to the singlet nitrene which does not discriminate between the disubstituted and the monosubstituted double bonds, the *triplet nitrene* prefers the *methyl substituted* double bond by a factor of 2 [87].

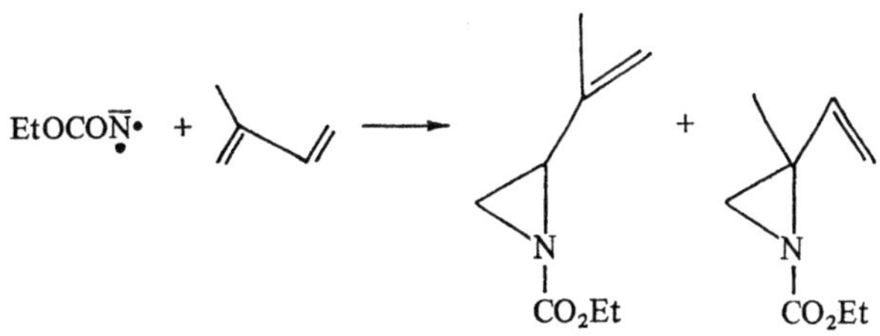

29' 43 44 45

With sulfonylnitrenes, additions to double bonds are very rare. The reason is that addition of the sulfonylazide to the double bond normally is much faster than azide decomposition. Another explanation is double-bond participation in the nitrogen elimination where no free nitrene is involved.

Only in the presence of metals, as in the copper-catalyzed decomposition of benzenesulphonylazide in cyclohexene [88] and in the photolysis of ferrocenylsulphonylazide in cyclohexane [10] such reactions are observed.

Cyanogenazide *46* reacts, unlike most organic azides, with strained and unstrained double bonds at room temperature to give alkylidenecyanamides *47* and/or N-cyanoazirines *48* [34,91]. These products are also expected from the reaction of cyanonitrene with olefins.

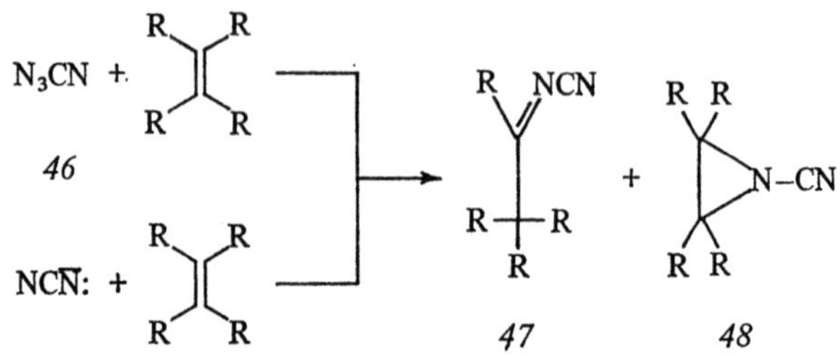

Therefore, without isotopic labeling, it is impossible to distinguish between azide-products and nitrene-products. That is the reason why only one NCN-olefin reaction (the thermolysis of NCN_3 in cyclooctatetraene) was studied in detail. Cyclooctatetraene (*49*) reacts with NCN_3 very slowly at room temperature and leads exclusively to the alkylidenecyanamide. Thermolysis of NCN_3 in diluted (ethylacetate) cyclooctatetraene at 78 °C afforded *50, 51, 52* [92,93].

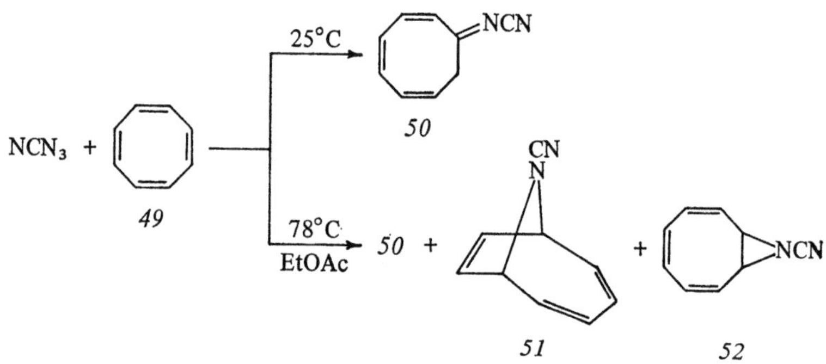

52 is formed in a one-step reaction by the singlet NCN̈ and *51* in a two-step reaction by the triplet NCN̄·. This explanation is supported by dilution experiments.

3.3. Dilution Experiments Effecting Intersystem Crossing to Triplet Nitrenes

As pointed out earlier, photolysis or thermolysis of organic azides generate the corresponding nitrenes initially in their singlet states. The triplet state must therefore be populated by intersystem crossing (isc). The dilution technique uses the different concentration dependence of the different steps involved. While isc of singlet to triplet nitrene is unimolecular, the [1+2]-cycloaddition step of the nitrene is first order in concentration. High dilution should therefore lead theoretically to the same cis-/trans-aziridine ratio. If the solvent used possesses a heavy atom (*e.g.* Br) then a fast intersystem-crossing step is effected; this is called *heavy atom effect.*

Thermolysis of ethylazidoformate *29* in cis-4-methylpent-2-ene *37* solutions in CH_2Cl_2 demonstrated that the ratio of cis- to trans-addition products *38/39* depends on the olefin concentration [86] (see Table 6).

Table 6. Thermolysis of ethyl azidoformate in cis-4-methylpent-2-ene solutions in dichloromethane at 100 °C for 24 h [86]

Olefin concn. mole [%]	Total aziridine [%]	Trans-aziridine [%]	[T]	[S]	[T]/[S]
100	67	14	0.175	0.825	0.21
33	—	18	0.225	0.775	0.29
10	78	33	0.410	0.590	0.70
5	—	41	0.515	0.485	1.06
3.3	60	46	0.575	0.425	1.35
1.5	—	51	0.64	0.36	1.78
1.0	—	54	0.675	0.325	2.08

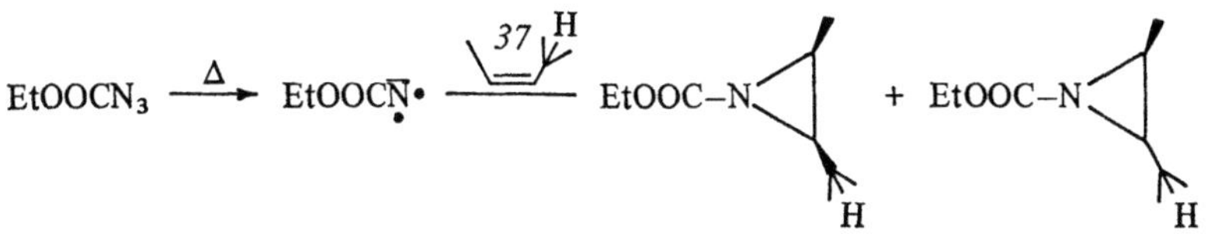

29 *29'* *38* *39*

The approximately comparable overall yield shows that the nonstereospecifity is increased by decreasing the olefine concentration. This is in agreement with the theory of collisional deactivation, which predicts that *dilution increases* the ratio of *triplet nitrene.* Thermolysis of NCN_3 in 1,2-dimethylcyclohexane *53* is another example of the dilution effect in nitrene reactions [94].

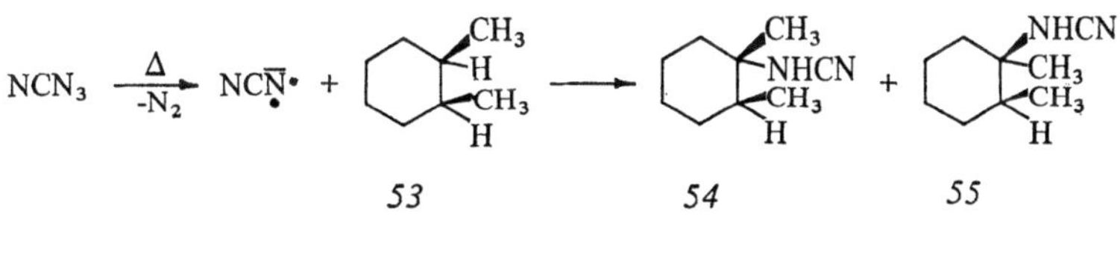

53 *54* *55*

Table 7. Stereochemistry of the insertion of cyanonitrene into the tertiary C—H bond of cis- and trans-1,2-dimethylcyclohexane as function of the solvent [94]

1,2-Dimethyl-cyclohexan	Solvent[1])	Temp.	% cis	% trans	Overall yield [%]
cis	—	46	>98	< 2	48
trans	—	46	< 2	>98	48
cis	CH_2Cl_2	41	74	26	28
trans	CH_2Cl_2	41	34	66	35
cis	CH_2Br_2	43	52	48	35
trans	CH_2Br_2	43	52	48	29

[1]) 10:1 solvent to hydrocarbon mixture.

From the data indicating partial loss of stereospecifity of the nitrene reaction in CH_2Cl_2 and the complete one in CH_2Br_2, it can be concluded that $NC\bar{N}\cdot$ reacts partly as triplet in CH_2Cl_2 (dilution effect) and exclusively as triplet in CH_2Br_2 (dilution and heavy atom effect). Similiar results were obtained in the thermolysis of NCN_3 in cyclooctatetraene *49* [93] (see Scheme p. 108).

Table 8. Relative amounts of *51* and *52* from reaction of $NC\bar{N}\cdot$ with cyclooctatetraene in various solvents [45]

Solvent	*51* [%]	*52* [%]	[*51*]/[*52*]
CH_2Br_2	76	24	3.3
$CH_3COOC_2H_5$	56	44	1.3
CH_3CN	17	83	0.2

In this case also the ratio *51*/*52* depends on the solvent. It is largest in CH_2Br_2, a solvent which is expected to favor reaction by $^3NC\bar{N}\cdot$, and smallest in CH_3CN, a solvent favoring reaction by $^1HC\bar{N}$:.

In contrast to the carbene-chemistry there is ample evidence that in some cases the singlet state of nitrene will be stabilized by inert solvents, *e.g.* C_6F_6 and CH_2Cl_2. The stabilization of nitrene singlet states by symmetrical interaction of the nitrene with two lone pairs has been proposed by Gleiter and Hoffman [95].

Breslow has reported that yields of insertion (singlet) of carbalkoxynitrene with cyclohexane are increased by dilution with hexafluorobenzene [96]. One of the explanations given for this fact is that the singlet nitrene complexes with the fluorine in C_6F_6 to give a nitrenoide-type species which is stable enough to decrease the rate of intersystem crossing but also active enough to insert in C—H bonds of saturated hydrocarbons. For a number of alkanoylnitrenes (RCO$\bar{N}$:) it was also shown that their singlet state is stabilized by CH_2Cl_2 [97]; and recently Belloli and co-workers [98] have shown that by dilution with CH_2Cl_2 carbalkoxynitrenes (ROCO$\bar{N}$:) singlets are also stabilized.

3.4. Triplet Nitrenes by Sensitization

Sensitized photolysis is a much more efficient method to produce triplet nitrenes than the collisional deactivation method. This technique, which populates the triplet state of a nitrene exclusively by energy transfer, employs triplet sensitizers such as aromatic ketones with a high intersystem-crossing efficiency. Energy is

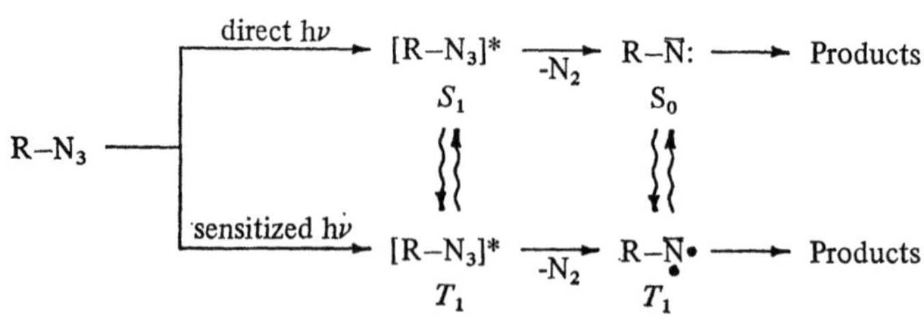

then transferred from the triplet sensitizer to the azide, populating the triplet state of the latter. The triplet azide then looses nitrogen and leads to a nitrene in the triplet state.

Not much data are available in this field. One example is the decomposition of hexanoylazide (*56*) in cyclohexane by direct and by sensitized photolysis [99].

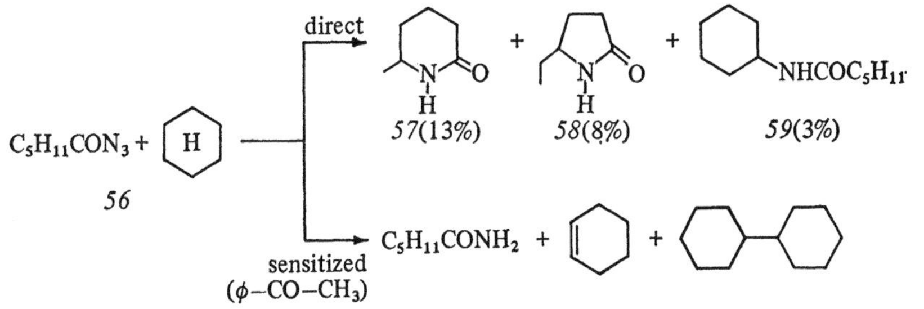

56; *57*(13%); *58*(8%); *59*(3%); *60*(78%); *34*(11%); *61*(11%)

In the direct photolysis the main products are isocyanates. Two lactames *57* and *58* and the insertion product *59* are formed in addition. The sensitized photolysis, in striking contrast to the direct photolysis, leads only to abstraction products *60, 34* and *61* in good yields arising from a triplet nitrene. No lactames could be detected here.

Another example is the photolysis of NCN_3 in cis-1,2-dimethyl-cyclohexane (*50*) [45].

Table 9. Stereochemistry of the insertion of photochemically generated cyanonitrene into tertiary C—H bonds of cis-1,2-dimethylcyclohexane [45]

Wavelength [A°]	Sensitizer	R—NHCN	
		% cis	% trans
3100—4500	—	>98	< 2
3100—4500	Fluorenone	87	13
3100—4500	Benzophenone	56	44

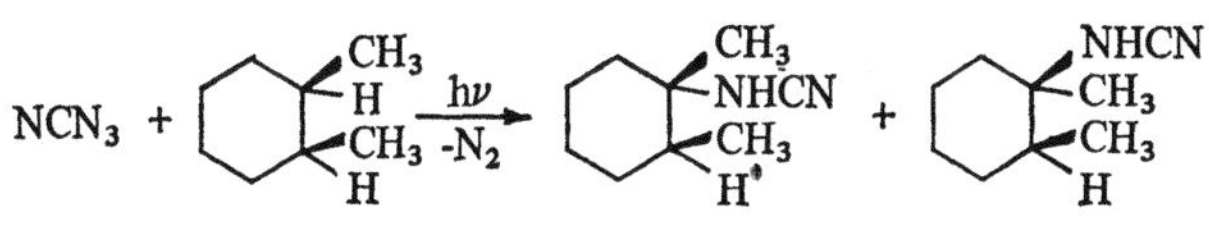

53 *54* *55*

Direct photolysis leads exclusively to the singlet nitrene which inserts into the tertiary C—H bond of *53* in a totally stereospecific fashion. In the sensitized photolysis C—H-insertion is no longer stereospecific; the loss of the stereospecifity parallels the efficiency of the sensitizer in promoting the decomposition of the azide.

4. References

1) Tiemann, F.: Chem. Ber. *24*, 4162 (1891).
2) Beckman, A. O., Dickinson, R. G.: J. Am. Chem. Soc. *50*, 1870 (1928).
3) Kirmse, W.: Angew. Chem. *71*, 537 (1959).
4) Horner, L., Christmann, A.: Angew. Chem. *75*, 707 (1963).
5) Abramovitch, R. A., Davis, B. A.: Chem. Rev. *64*, 149 (1964).
6) Lwowski, W.: Angew. Chem. Intern. Ed. Engl. *6*, 897 (1967).
7) Kirmse, W.: In: Carbene, carbenoide and carbenanaloge. Weinheim: Verlag Chemie 1969.
8) Gilchrist, T. L., Rees, C. W.: In: Carbenes, nitrenes and arynes. New York: Appleton-Century-Crofts 1969.
9a) Abramovitch, R. A.: Chem. Soc., Spec. Publ. *24*, 323 (1970).
9b) Abramovitch, R. A.: In: Organic reactive intermediates (S. P. Mc Manus, ed.). London: Academic Press 1973.
10) Abramovitch, R. A., Sutherland, R. G.: Fortschr. Chem. Forsch./Topics Curr. Chem. *16*, 1 (1970).
11) Lwowski, W. (ed.): Nitrenes. New York: John Wiley and Sons Inc. 1970.
12) Belloli, R.: J. Chem. Educ. *48*, 423 (1971).
13) Abramovitch, R. A., Kyba, E. P.: J. Am. Chem. Soc. *93*, 1537 (1971).
14) Lüttringhaus, A., Jander, J., Schneider, R.: Chem. Ber. *92*, 1756 (1959).
15) Rice, F. O., Luckenbach, T. A.: J. Am. Chem. Soc. *82*, 2681 (1960).
16) Heacock, J. F., Edmison, H. T.: J. Am. Chem. Soc. *82*, 3460 (1960).
17) Smolinsky, G.: J. Am. Chem. Soc. *83*, 2489 (1961).
18) Knunyants, J. L., Bykhovskaya, E. G.: Proc. Acad. Sci. USSR *132*, 513 (1960).
19) Kirmse, W.: Angew. Chem. *71*, 450 (1959).
20) Smolinsky, G.: J. Am. Chem. Soc. *82*, 4717 (1960).
21a) Ramsperger, H. C.: J. Am. Chem. Soc. *51*, 2134 (1929).
21b) Patai, S., Gotshal, Y.: J. Chem. Soc. B, *1966*, 489.
22) Saunders, W. H., Jr., Ware, J. C.: J. Am. Chem. Soc. *80*, 3328 (1958).
23) Walker, P., Waters, W. A.: J. Chem. Soc. *1962*, 1632.
24) Appl, M., Huisgen, R.: Chem. Ber. *92*, 2961 (1959).
25) Breslow, D. S., Sloan M. F., Newburg, N. R., Renfrow, W. B.: J. Am. Chem. Soc. *91*, 2273 (1969).
26) Huisgen, R., Blaschke, H.: Tetrahedron Letters *1964*, 1409.
27) Prosser, T. J., Marcantonio, A. F., Breslow, D. S.: Tetrahedron Letters *1964*, 2479.
28) Anastassiou, A. G., Simmons, H. E.: J. Am. Chem. Soc. *89*, 3177 (1967).
29) Reiser, A., Marley, R.: Trans. Faraday Soc. *64*, 1806 (1968).
30) Reiser, A., Bowes, G., Horne, R. J.: Trans. Faraday Soc. *62*, 3162 (1966).
31) Lewis, F. D., Saunders, W. H.: J. Am. Chem. Soc. *90*, 3828 (1968). — Lewis, F. D., Saunders, W. H.: J. Am. Chem. Soc. *90*, 7031 (1968).
32) Lewis, F. D., Saunders, W. H.: J. Am. Chem. Soc. *90*, 7033 (1968).
33) Closson, W. D., Gray, H. B.: J. Am. Chem. Soc. *85*, 290 (1963).
34) Marsh, F. D., Hermes, M. E.: J. Am. Chem. Soc. *86*, 4506 (1964).
35) Lwowski, W., Mattingly, W., Jr.: J. Am. Chem. Soc. *87*, 1947 (1965).
36) Horner, L., Christmann, A.: Chem. Ber. *96*, 388 (1963).
37) Reagan, M. T., Nickon, A.: J. Am. Chem. Soc. *90*, 4096 (1968).
38a) For a review, see Berry, R. S.: In: Nitrenes (W. Lwowski, ed.). New York: Wiley-Interscience 1970.
38b) Huo, W. M.: J. Chem. Phys. *49*, 1482 (1968).
39) Higuchi, J.: J. Chem. Phys. *24*, 535 (1956).
40) Ransil, B. J.: Rev. Mod. Phys. *32*, 245 (1960).
41) Hurley, A. C.: Proc. Roy. Soc. (London) *A 248*, 119 (1958).
42) Boyd, M. E.: J. Chem. Phys. *29*, 108 (1958).
43) Krauss, M., Wehner, F.: J. Chem. Phys. *29*, 1287 (1958).
44) Yarkony, D. R., Schaefer, H. F., Rothenberg, S.: J. Am. Chem. Soc. *96*, 5974 (1974).

45) Anastassiou, A. G., Shepelavy, J. N., Simmons, H. E., Marsh, F. D.: In: Nitrenes (W. Lwowski, ed.). New York: Wiley-Interscience 1970.
46) Alewood, P. F., Kazmaier, P. M., Rank, A.: J. Am. Chem. Soc. *95*, 5466 (1973).
47) Smolinsky, G., Wasserman, E., Yager, W. A.: J. Am. Chem. Soc. *84*, 3220 (1962).
48a) Barash, L., Wasserman, E., Yager, W. A.: J. Am. Chem. Soc. *89*, 3931 (1967).
48b) Wasserman, E., Smolinski, G., Yager, W. A.: J. Am. Chem. Soc. *86*, 3166 (1964).
49) Wasserman, E., Barash, L., Yager, W. A.: J. Am. Chem. Soc. *87*, 2075 (1965).
50) Wasserman, E., Snyder, L. C., Yager, W. A.: J. Chem. Phys. *41*, 1763 (1964).
51) Reiser, A., Frazer, V.: Nature *208*, 682 (1965).
52) Reiser, A., Wagner, H., Bowes, G.: Tetrahedron Letters *1966*, 2635.
53) Reiser, A., Terry, G. C., Willets, F. W.: Nature *211*, 410 (1966).
54) Reiser, A., Wagner, H. M., Marley, R., Bowes, G.: Trans. Faraday Soc. *63*, 2403 (1967).
55) Milligan, D. E., Jacox, M. E., Comeford, J. J., Mann, D. E.: J. Chem. Phys. *43*, 756 (1965).
56) Milligan, D. E., Jacox, M. E., Bass, A. M.: J. Chem. Phys. *43*, 3149 (1965).
57) Breslow, D. S., Edwards, E. J.: Tetrahedron Letters *1967*, 2123.
58) Pritzkow, W., Timm, D.: J. Prakt. Chem. *32*, 178 (1966).
59) Moriarty, R. M., Rahman, M.: Tetrahedron *21*, 2877 (1965).
60) Leermakers, J. A.: J. Am. Chem. Soc. *55*, 2719 and 3098 (1933).
61) Bertho, A.: Chem. Ber. *57*, 1138 (1934).
62) Smith, P. A. S., Hall, J. H.: J. Am. Chem. Soc. *84*, 480 (1962).
63) Hall, J. H., Hill, J. W., Tsai, H. C.: Tetrahedron Letters *1965*, 2211.
64) Smolinsky, G.: J. Org. Chem. *26*, 4108 (1961).
65) Smith, P. A. S.: In: Nitrenes (W. Lwowski, ed.). New York: John Wiley and Sons Inc. 1970.
66) Hall, J. H., Hill, J. W., Fargher, F.: J. Am. Chem. Soc. *90*, 5313 (1968); for further Ref. see also [65]).
67) Abramovitch, R. A., Azogu, C. I., McMaster, J. T.: J. Am. Chem. Soc. *91*, 1219 (1969).
68) Abramovitch, R. A., Roy, J., Uma, V.: Can. J. Chem. *43*, 3407 (1965).
69) Sloan, M. F., Breslow, D. S., Renfrow, W. B.: Tetrahedron Letters *1964*, 2905.
70) Dermer, O. C., Edmison, M. T.: J. Am. Chem. Soc. *77*, 70 (1955).
71) Tilney-Bassett, J. F.: J. Chem. Soc. *1962*, 2517.
72) Abramovitch, R. A., Knaus, G. N., Uma, V.: J. Org. Chem. *39*, 1101 (1974).
73) Breslow, D. S., Prosser, T. J., Marcantonio, A. F., Genge, C. A.: J. Am. Chem. Soc. *89*, 2384 (1967).
74a) Skell, P. S., Woodworth, R. C.: J. Am. Chem. Soc. *78*, 4496 (1956); *81*, 3383 (1959).
74b) Hoffmann, R.: J. Am. Chem. Soc. *90*, 1475 (1968).
75) Fagley, T. F., Setter, J. R., Oglukian, R. L.: J. Am. Chem. Soc. *78*, 5567 (1956).
76) Russel, K. E.: J. Am. Chem. Soc. *77*, 3487 (1955).
77) Walker, P., Waters, W. A.: J. Chem. Soc. *84*, 480 (1962).
78) Scheiner, P.: J. Am. Chem. Soc. *90*, 988 (1968).
79) Lwowski, W., Maricich, T. J.: J. Am. Chem. Soc. *87*, 3630 (1965).
80) Lwowski, W., McConaghy, J. S., Jr.: J. Am. Chem. Soc. *87*, 5490 (1965).
81) Lwowski, W., Woerner, F. P.: J. Am. Chem. Soc. *87*, 5491 (1965).
82) Lwowski, W., Mattingly, T. W., Jr.: Tetrahedron Letters *1962*, 277.
83) McConaghy, J. S., Jr., Lwowski, W.: J. Am. Chem. Soc. *89*, 2357 (1967).
84) Berry, R. S., Cornell, D. W., Lwowski, W.: J. Am. Chem. Soc. *85*, 1199 (1963).
85) Cornell, D. W., Berry, R. S., Lwowski, W.: J. Am. Chem. Soc. *87*, 3626 (1965).
86) McConaghy, J. S., Jr., Lwowski, W.: J. Am. Chem. Soc. *89*, 4450 (1967).
87) Mishra, A., Rice, S. N., Lwowski, W.: J. Org. Chem. *33*, 481 (1968).
88) Abramovitch, R. A., Azogu, C. D., Sutherland, R. G.: Chem. Commun. *1969*, 1439.
89) Franz, J. E., Osuch, C., Dietrich, M. W.: J. Org. Chem. *29*, 2922 (1964).
90) Oelschlager, A. C., Zalkow, L. H.: Chem. Commun. *1966*, 5.
91) Anastassiou, A. G.: J. Org. Chem. *31*, 1131 (1966).
92) Anastassiou, A. G.: J. Am. Chem. Soc. *87*, 5512 (1965).
93) Anastassiou, A. G.: J. Am. Chem. Soc. *90*, 1527 (1968).
94) Anastassiou, A. G.: J. Am. Chem. Soc. *88*, 2322 (1966).
95) Gleiter, R., Hoffman, R.: Tetrahedron *24*, 5899 (1968).

[96] Breslow, D. S., Edwards, E. I.: Tetrahedron Letters *1972*, 2041.
[97] Felt, G. R., Linke, S., Lwowski, W.: Tetrahedron Letters *1972*, 2037.
[98] Belloli, R., Whitehead, M. A., Wollenberg, R. H., La Bahn, V. A.: J. Org. Chem. *39*, 2128 (1974).
[99] Brown, J., Edwards, O. E.: Can. J. Chem. *45*, 2599 (1967).

Received February 3, 1975

Spectroscopic Implications of Line Broadening in Large Molecules

Professor Gad Fischer

Department of Chemistry, Ben-Gurion University of the Negev, Beer-Sheva 84120 Israel

Contents

1. Introduction

A discussion on line broadening in the spectra of large molecules may be best introduced by describing the spectral results obtained for naphthalene when present as a low-concentration substitutional solid solution with durene at very low temperatures [1,2].

It may be considered surprising that the triplet–triplet absorption spectrum [2], Fig. 1, displays a series of featureless broad bands (width at half-height $\sim$300 cm^{-1}) whereas the lowest singlet–singlet transition of the same sample under identical physical conditions gives bands two orders of magnitude sharper. On the other hand, for transitions to the higher-lying singlet states, increased band widths are obtained. The immediate retort may be that one is not really observing a molecular characteristic but an electronic transition in which the entire mixed crystal is involved. Closer inspection of the effects of environmental broadening reveals that, although it is not possible to reject the host crystal outright as being responsible for the selective broadening in the triplet–triplet and higher singlet systems, it is difficult to reconcile the range of broadenings observed with crystal effects [3].

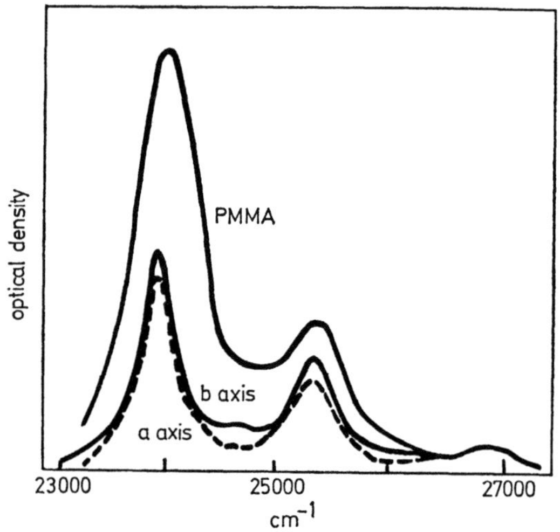

Fig. 1. Triplet-triplet absorption spectra of naphthalene in polymethylmethacrylate at 293 K (PMMA) and parallel to the *a* and *b* axes of durene host crystal at 4 K

Similar results have been reported for naphthalene in the vapor phase by Porter and Wright [4]. However, in the vapor phase at room temperature the observed broadenings may result from spectral congestion. Each vibronic band is associated with extensive rotational structure and in addition sequence and hot bands are active. Furthermore, there is no reason why exactly similar spectral congestion must obtain for different electronic transitions.

One is therefore led to look for essential differences between the varying electronic transitions, which may be reflected in the varying linewidths. As the respective transitions are all spin and orbitally (electric dipole) allowed, this aspect of the transition is not a contributory factor, although transition intensities may be important. There are two clear-cut differences, and they concern

(i) the absence of intermediate singlet states in the energy interval between the electronic ground and the lowest excited singlet state. This is not the case for transitions to higher-lying singlet states nor does a corresponding situation prevail for the triplet–triplet transition.

(ii) the energetics, the final triplet state, and likewise the higher singlets are located at much higher energies relative to the ground state than is the first excited singlet.

If these two observations are indeed pertinent, this would suggest that, under certain circumstances, tansitions between the two lowest triplet states may be sharp. Of course, it would be necessary to choose a system where the two lowest triplet states are separated by an energy gap typical of the separation of the ground and first excited singlet states.

One further observation that is very relevant to any comparison of the characteristics of the various singlet–singlet and triplet–triplet transitions concerns the general absence under typical experimental conditions of fluorescence from other than the lowest excited singlet state, no triplet–triplet fluorescence being recorded, and the occurrence of phosphorescence being recorded only from the lowest triplet state.

It is the purpose of this review to consider in detail the implications of the above observations on the energetics of the absorbing states, the presence of intermediate electronic states, and emission quantum yields vis-à-vis line broadening in the spectra of large molecules. In the following section we consider the reported contributions to the explanation of the possible causes of diffuseness in electronic spectra. We also define what is conventionally understood by large molecules and line broadening. Numerous widely representative, experimentally recorded spectra are presented in Section 3. In the final section we consider the theoretical models that have been proposed to account for the observed spectra. Despite the difficulties involved in carrying out actual calculations for real systems, progress has been made in this direction.

2. Diffuseness in Electronic Spectra

2.1. General

Byrne and Ross [5] have considered diffuseness in electronic spectra and have listed some twelve causes of line broadening. In an earlier work [6] they considered in detail a trivial explanation of the broadening, namely spectral congestion, with particular reference to the related molecules benzene, naphthalene, anthracene and tetracene. They showed that for the first three molecules spectral fine structure should be observable under the appropriate experimental conditions, but that for tetracene under practical experimental conditions no resolvable fine structure is predicted. They supported these conclusions by recording the vapor absorption spectrum of anthracene, the fine structure of which was clearly resolvable.

All of Byrne and Ross' work referred to the vapor-phase spectra. In a complementary work, Hochstrasser and Marzzacco [3] and Hochstrasser [7] considered the perturbations between the low-lying electronic states of a number of aromatic and heteroaromatic molecules in the solid state. They recorded the low-temperature electronic spectra, and suggested that the appearance of line broadening could be viewed as an indication of the presence of another electronic state. This is of particular relevance for the aza-aromatic molecules, since for them there is an increased number of low-lying electronic states, due to the presence of the nitrogen lone-pair electrons. Hochstrasser and Marzzacco further concluded that interactions between $\pi\pi^*$ and $n\pi^*$ states are larger than interactions between states of the same orbital type, such as $\pi\pi^*$ by $\pi\pi^*$, and this conclusion was supported by experimental observations on the line broadenings of the higher-lying excited electronic states. In this context, the lowest excited electronic state of each multiplicity was excluded from consideration.

In a review on radiationless transitions Jortner, Rice and Hochstrasser [8] described some results from solid-state spectra. The two most important generalizations are:

i) Transitions to higher-lying states of the one multiplicity display broader bands than those to the the first excited state. This characteristic is not an outcome of uncertainty broadening due to radiative effects.

ii) The broadening is dependent on the energy gaps separating the state in question and lower-lying states of the same multiplicity.

Other general results appearing from the experimental study concern the size of the molecule, the oscillator strength of the transition, and the nature of the transition, such as whether it is $n\pi^*$, $\pi\pi^*$, or Rydberg. In their review on diffuseness in gas-phase spectra they emphasized the importance of first establishing the nature of the broadening mechanism. When it is possible to eliminate trivial (extrinsic) causes such as spectral congestion and Döppler broadening, the same conclusions are reached as are listed above for the solid state spectra.

Apart from the above-mentioned studies of Hochstrasser and co-workers, and Ross and co-workers on linewidths in vapor-phase and solid-state spectra, carried out in the last few years little attention has been paid to the detailed study of linewidths in the electronic spectra of large molecules. It is not inappropriate to remark that, until recently, there has been little or no interest in transitions to higher-lying excited electronic states, excluding the Rydbergs, for the precise reason that in general only broadened bands were observed. All the attention was therefore devoted to the lowest singlet and triplet systems since they displayed a sharp structure for which, in most cases, it was possible to arrive at consistent vibrational analyses, and in some cases also rotational analyses. In addition, fluorescence and phosphorescence studies indicated, almost without exception, that only the lowest excited state was capable of emitting. Hence it was concluded that the lifetime of the molecule in its higher excited states was considerably shorter than its natural radiative lifetime as estimated from the absorption osciallator strength.

It is therefore clear that, although the occurrence of rapid electronic relaxation from higher-lying electronic states has been recognized for some time now, the relation between it and line broadening has only recently been emphasized.

In any study of the line broadening in large molecules it is important to specify what is understood by large molecules and to define line broadening. Large molecules are molecules which, for energy gaps of the order of 10^4 cm^{-1}, satisfy the physical conditions pertaining to the statistical case. Since the discussion is restricted mainly to aromatic and aza-aromatic molecules, the smallest molecules we shall consider here are benzene and sym-triazine. In terms of the electronic spectroscopy of polyatomic molecules these are generally considered large molecules, since individual rotational lines cannot be identified, in contrast to the small molecules such as SO_2, NSF, and H_2CO. Furthermore, for these small molecules, transitions to the higher electronic states may, and often do, display bands as sharp as to the lowest excited state.

In the definition of line broadening it is necessary to exercise some discrimination. On the one hand spectral linewidths of less than 0.17 cm^{-1} are observed for some of the vibronic bands of the lowest singlet system of benzene $^1B_{2u} \rightarrow {}^1A_{1g}$ in the vapor phase [9], while on the other hand many electronic spectra have been encountered, in particular in higher excited singlet and triplet systems, for which few or no vibrational features are apparent. In crystal spectra at 4 K, linewidths as sharp as 0.5 cm^{-1} are often obtained for the lowest excited state of any multiplicity, despite coupling with the lattice modes, which may be expected to lead to considerable broadening. Nevertheless, these crystal linewidths are considerably more than the linewidths observed in the vapor phase and certainly more than the natural radiative widths.

In general, in any spectrum there are a number of contributions to the linewidth. If the purpose of a spectroscopic study concerns the assessment of a particular broadening mechanism, then it is essential that the other contributions to the broadening be held fixed or at least that their contributions are known while certain parameters, for example excitation energy, are varied.

2.2. Causes of Line Broadening

In vapor-phase spectra at low pressures (<10 torr) no problems of environmental broadening, collisions and intermolecular interaction are encountered. For the majority of aromatics and aza-aromatics, solids or liquids at room temperature, only very low pressures are achieved at 20 °C and in the experimentally practical temperature range 273 to 373 K.

Döppler widths given by

$$\Delta\nu = 2\frac{v}{c}\nu\,,$$

where v is the mean velocity of the absorber and ν is the absorption frequency in the absence of motion, only become a problem when the instrumental resolving power is sufficiently high. This was the case in the benzene study of Callomon,

Parkin and Lopez-Delgado, where resolving powers of ~750,000 were attained. However, the majority of large-molecule spectra have been recorded at resolving powers of less than the Döppler widths.

Spectral congestion is an important source of line broadening. Temperature-dependent studies may be very helpful in delineating the nature of the broadening mechanism. This is of particular significance in identifying sequence congestion. When such studies are taken together with solid-state studies at 4 K of the electronic origin and the accompanying low-lying vibrational bands, sequence diffuseness, rotational diffuseness and vibrational congestion may all be identified and eliminated. In the solid state, environmental broadening must be considered. This may arise through coupling with the lattice vibrational modes, through the ocurrence of a multiplicity of sites, or through Davydov splitting.

All of the above-mentioned causes of line broadening are trivial (extrinsic) in the sense that, given the appropriate experimental conditions, the cause of the line broadening may be removed or at least reduced to a minimum. There is no sure general prescription for selecting the appropriate experimental conditions but dilute mixed-crystal studies at 4 K to examine the spectrum in a variety of crystal hosts have proved successful in the majority of cases. The anthracene lowest singlet spectrum is an excellent example. The vapor phase offers a very narrow range of physical conditions so that it is possible to observe resolvable structure, while in host crystals of *p*-terphenyl bandwidths of ~1 cm^{-1} are obtained [10)].

Here we are chiefly interested in the intrinsic causes of line broadening. We do not include among these dissociation, ionization, predissociation, autoionization, and preisomerization, since few unambiguous examples of their occurrence have been reported for the low-lying excited electronic states. Attention is devoted to broadening through anharmonicity (vibrational relaxation) and in particular to electronic relaxation.

2.3. Anharmonicity

What evidence is there to support Neporent's [11)] arguments that intramolecular vibrational relaxations may generate diffuse and complex spectra in polyatomic molecules? Neporent does stress the point that the differentiation between electronic and vibrational relaxation is based on the Born-Oppenheimer approximation, and if the Born-Oppenheimer approximation is not tenable, as may be the case for the higher-lying electronic states, no differentiation is possible.

By line broadening through anharmonicity we understand a coupling responsible for the rapid redistribution of vibrational energy within the one electronic state. It is difficult to understand how a sufficiently strong coupling of the vibrational levels of the lower-lying electronic states may arise. In the usual approach, anharmonicity is associated with the terms of higher order than quadratic in the expansion of the electronic potential about some fixed nuclear configuration. But to achieve the coupling of a large number of energetically close-lying levels at high vibrational energies, the low-order cubic and quartic terms are not sufficient. The cubic and quartic terms are too selective to bring about a coupling of all the levels since vibrational matrix elements of the type

$$<\chi_i(Q)\,|Q_a^{n_a}\,Q_b^{n_b}\ldots|\,\chi_j(Q)>$$

must be non-vanishing and the coupling terms $Q_a^{n_a}\,Q_b^{n_b}\ldots$ must be totally symmetric. χ_i are the vibrational wave functions, functions of the normal coordinates Q, and n_a is the power to which the normal coordinate Q_a appears in the coupling terms. Instead, higher-order terms are required. However, the magnitude of the coupling is dependent on the order, and for high-order terms it is very small. Of course, if the vibrational energies are approaching the dissociation energy, the harmonic description of the electronic potential energy as a function of normal coordinates is no longer appropriate, and the magnitude of the coupling through higher-order terms may become very significant.

Experimentally, there is little to suggest that large anharmonicities may be expected for the low-lying electronic states of the aromatic and aza-aromatic molecules, at least on the evidence of the extensive sequence structure and the well-developed progressions observed for the lowest excited singlet and triplet states of many molecules.

2.4. Electronic Relaxation

Among the numerous intrinsic causes of line broadening in the electronic spectra of large molecules, electronic relaxation (widely called radiationless transitions) is the most important and has received the greatest attention in recent times.

In the model generally employed to describe the line broadening it is assumed that there exist two sets of interacting levels corresponding to the Born-Oppenheimer levels of two different electronic states, where one of the electronic states contains only one level ψ_n of energy E_n, to which optical transitions are allowed from the ground state [12]. The lineshape is given by a Lorentzian function of width

$$\Delta = 2\pi\, V^2 \varrho\,. \tag{1}$$

In this simplified model, later to be discussed in more detail, the linewidth depends directly on $V^2\varrho$ and, clearly, the greater this product the greater the linewidth. It should be noted at the outset that in this unsophisticated model the quantity ϱV^2 is assumed to be constant, although it is known that ϱ, the density of states, is a rapidly increasing function of E and that V, the vibronic coupling, varies over many orders of magnitude and also varies in sign [12,13]. However, despite these variations, it is possible to consider an average value of the quantity ϱV^2 and to predict an average lineshape. This stems from the fact that the sign of V is unimportant since in the model of one discrete, optically active state interacting with a dense manifold which does not carry intensity, only the squared interaction is considered [14].

In order to obtain the Lorentzian lineshape, that corresponds to the statistical limit, it is necessary that the average vibronic coupling matrix element is much larger than the mean spacing between the levels. In terms of the linewidth Δ, it is clear that $\Delta \gg \varepsilon$ if we are to speak of a Lorentzian line-shape, which in turn requires that $(V\varrho)^2 \gg 1$.

When we consider the broadening in terms of the uncertainty principle,

$$\Delta\nu\,\tau \geqslant \frac{1}{4\pi c},$$

where $\Delta\nu$ is a measure of the uncertainty broadening in cm^{-1}. τ contains contributions from both the radiative K_r and radiationsless K_{nr} channels

$$\tau = (K_r + K_{nr})^{-1} \tag{2}$$

where K_{nr} and K_r are the radiationless and radiative rate constants, respectively. In the statistical limit $K_{nr} \gg K_r$ and hence

$$\tau \simeq K_{nr}^{-1}. \tag{3}$$

A distinction arises in the description of the lineshape when the dense, optically inactive manifold is replaced by a sparse one, corresponding to the intermediate case [15]. This may apply to small molecules such as SO_2 and NO_2, and also to large molecules characterized by a small energy gap between the two interacting electronic states, such as naphthalene [16,17] and 3,4-benzpyrene [18]. According to Jortner, Rice and Hochstrasser [8], separate, inhomogenously broadened lines are predicted for the intermediate case.

3. Spectral Observations

3.1. Benzene

The benzene spectrum is one of the most studied and one of the best-known illustrations of the occurrence of line broadening in electronic tiansitions to the higher-lying excited states of a large molecule. The 2600 ($^1B_{2u} \leftarrow {}^1A_{1g}$) and 2000 Å ($^1B_{1u} \leftarrow {}^1A_{1g}$) systems, observed under low resolution in the vapor phase, are displayed in Fig. 2 [19]. It is apparent that, despite the low spectral resolution, bandwidths in the latter system are very considerably greater. This relative characteristic is not lost in the low-temperature solid-state spectra where in the crystal at 4 K linewidths as small as 1 cm^{-1} are obtained for the first singlet system [20], compared to greater than 100 cm^{-1} for the second system [21,22]. In a krypton matrix linewidths of 10—35, 300 and ~350 cm^{-1} are recorded for the first three $\pi\pi^*$ systems, respectively [23].

Callomon, Parkin and Lopez-Delgado [9] have studied the variation in line widths with increasing vibrational energy in the lowest singlet system of benzene in the vapor phase. They took as a lower limit of detectable diffuseness $\Delta\bar{\nu} \simeq$ 0.17 cm^{-1}, where $\Delta\bar{\nu}$ is the spectral linewidth. More accurate estimates of the dif-

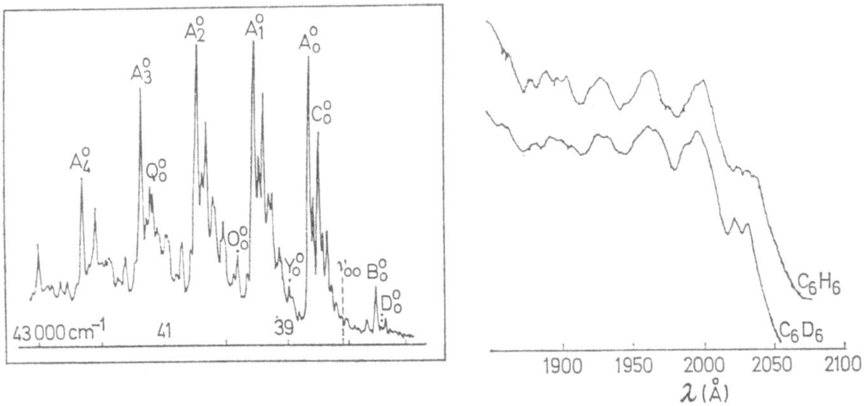

Fig. 2. The 2600 and 2000 Å absorption spectra of benzene vapor, according to Dunn [19]

fuse linewidths can be made by comparing observed with calculated rotational contours in which the linewidth is a widely varying parameter. Callomon, Parkin and Lopez-Delgado indicate that the contours are very sensitive to line broadening in the region of particular interest, $\Delta\bar{\nu} \simeq 0.05$ to $0.2\ \mathrm{cm}^{-1}$. Fluorescence is also detected from the low-lying vibrational levels of this electronic state indicating that the ratio of the radiative to the radiationless rate for these levels must be greater than 1:300. On the other hand the spectral linewidths foı the ${}^1B_{1u} \leftarrow {}^1A_{1g}$ and ${}^1E_{1u} \leftarrow {}^1A_{1g}$ systems under optimum expeıimental conditions are much greater. In the vapor phase the latter appears with very little vibrational structure [19].

This description of the relative spectral linewidths of the lowest excited $\pi\pi^*$ states applies to the whole family of aromatic hydrocarbons. It also applies to the manifold of triplet $\pi\pi^*$ states. In the case of benzene, Burland, Castro and Robinson [24] and Burland and Castro [25] have used phosphorescence and delayed fluorescence excitation techniques, respectively, to measure the absorption spectrum of the lowest triplet state, ${}^3B_{1u}$ of ultrapure crystals at 4 K. The origin is located at 29647 cm^{-1}. Unlike all the earlier studies on the lowest singlet → triplet absorption spectrum, this was not an O_2 perturbation experiment. Here widths of less than 3 cm^{-1} were obtained. This result should be compared with the much broader bands $\sim 150^{-1}$ observed for the suspected second triplet ${}^3E_{1u}$ in 5 cm crystals of highly purified benzene [26]. The two triplet states are separated by 7300 cm^{-1}.

The particular mechanism of electronic relaxation for the ${}^1B_{1u}$ state is not well understood. Shimakura, Fujimuıa and Nakajima [27] have calculated the nonradiative transition rate for the ${}^1B_{1u} \rightarrow {}^1B_{2u}$ transition by taking into account the pseudo Jahn-Teller distortion in the ${}^1B_{1u}$ state through vibronic coupling with the electronically degenerate ${}^1E_{1u}$ state. For the ${}^1B_{1u}$ state of a molecule belonging to D_{6h} point group symmetry, the direct radiationless transition ${}^1B_{1u} \rightarrow {}^1B_{2u}$ vanishes because the transition density between these two states is exactly zero in the π-electron approximation. The matrix element

$$\left\langle \Phi^0(^1B_{1u}) \left| \left(\frac{\partial U(qQ)}{\partial Q_{a_{2g}}}\right)_{Q_0} \right| \Phi^0(^1B_{2u}) \right\rangle = 0 \tag{4a}$$

but

$$\left\langle \Phi^0(^1B_{1u}) \left| \left(\frac{\partial U(qQ)}{\partial Q_{e_{2g}}}\right)_{Q_0} \right| \Phi^0(E^{\pm}_{1u}) \right\rangle \neq 0 \tag{4b}$$

These authors arrived at a calculated half-width of 150 cm^{-1}, in good agreement with experiment. However, their calculation did not take into account the presence of the as yet unidentified electronic relaxation channel observed by Callomon, Parkin and Lopez-Delgado [9] to be predominant in the radiationless transition from the higher vibrational levels of the $^1B_{2u}$ state. This channel may contribute to the broadening in the $^1B_{1u}$ state. Further extensive photodissociation is observed upon excitation into the $^1B_{1u}$ state and therefore photodissociation must be considered a possible cause of the line broadening.

In a high-resolution study of the linewidths in the 2600 Å system of benzene, Callomon, Parkin and Lopez-Delgado noted that the majority of bands involving vibrational levels more than 3000 cm^{-1} above the zero-point level of the excited state are diffuse. They pointed out that this threshold agrees with the cut-off in the excitation of fluorescence. No resonance fluorescence can be excited by excitation to vibrational levels lying above this threshold. The evidence suggested that a third and as yet undefined relaxation channel became operative at energies greater than 3000 cm^{-1} above the zero-point energy in the $^1B_{2u}$ state. The absence of resonance fluorescence at these higher energies indicated that the radiationless rate was now much greater than the radiative rate, at least by a factor of 300 to 1. This set a minimum for the nonradiative transition rate for this new channel of $K_{nr} \geqslant 10^9$ sec^{-1}. The high-resolution linewidth studies established a relaxation rate of $K_{nr} \simeq 3 \times 10^{10}$ sec^{-1}. There remains the intriguing question of identifying this channel. The absence of measureable quantities of photodecomposition products rules out photochemical processes. Assignment of this channel to an unidentified electronic state appears to offer the most promise.

3.2. Small-Molecule Behavior

(a) *Naphthalene and 1,6-Naphthyridine*

For the situation where the energy gap between the two interacting electronic states is small, *i.e.* of the order of vibrational frequencies, small-molecule behavior may be observed [15]. This is the recorded case for naphthalene [17], pyrene [28], biacetyl [29], 3,4-benzpyrene [18] and benzophenone [30,31] to name just a few examples. The electronic relaxation may be aptly described by the Intermediate Case, which is characterized by the following features. Nonradiative electronic relaxation does not occur in the isolated molecule. The density of vibronic states is small relative to the statistical situation while the vibronic coupling matrix elements are large. The coarse spacing of levels exceeds the radiative bandwidth resulting in a spectrum of well-separated lines exhibiting the strongly coupled sparsely spaced $\{\phi_l\}$ levels that borrow intensity from the ϕ_s state [15,17].

In the naphthalene vapor spectrum, the combination of extensive sequence structure, extending for the S_1 ($^1B_{2u}$) system more than 200 cm^{-1} to the red of the parent band, and the broad rotational contour obscure direct measurements of the intrinsic line widths [32]. In the durene mixed crystal spectra at 4 K linewidths of ~ 1 cm^{-1} were observed for the first system [16]. In the same host crystal, the second excited state of naphthalene, $^1B_{1u}$ short axis polarized, lies only ~ 2500 cm^{-1} above the $^1B_{2u}$ long axis polarized state. In the region of overlap a very complex spectrum is obtained, which has only recently been analyzed [16]. Considerable vibrational structure is apparent and both sharp and broad bands are noted. The spectrum observed is not merely a superposition of the two separate and noninteracting spectra but instead one recording transitions to a mixed state. The two lowest excited singlet states are of different symmetries and in a durene crystal the transitions appear to be polarized parallel to different crystallographic axes: a/b and c'. Relative to the two noninteracting states, significant level shifts and intensity transfers have occurred.

Very similar spectral results are obtained from pyrene in mixed crystals of biphenyl and fluorene at about 10 K [28]. For pyrene the energy gap between the $^1B_{2u}$ and $^1B_{1u}$ origins is approximately 2000 cm^{-1} in fluorene and 1700 cm^{-1} in biphenyl crystals. These values are less than for the corresponding gap of naphthalene. However, the increased number of normal coordinates can be expected to lead to a similar density of states, and in fact a range of spectral linewidths paralleling those obtained for naphthalene is recorded.

For a number of aza-naphthalenes similar observations concerning the spectra of two close-lying electronic states have been reported [33]. The interesting and real possibility arises that the interacting electronic states are of different orbital symmetry, in particular $n\pi^*$ and $\pi\pi^*$ states, in contrast to naphthalene and pyrene where the relevant states are both $\pi\pi^*$. As the interactions between $n\pi^*$ and $\pi\pi^*$ states are expected to be much larger than interactions between two $\pi\pi^*$ or two $n\pi^*$ states [7], the effects of the increased interaction may be apparent in the spectra. However this may not be the case when the two interacting electronic states are close-lying. The electronic matrix elements do not show any systematic variation with the nature of the combining states [3].

The coupling V in Eq. (1) may be written as a product of a vibronic coupling energy and a vibrational overlap integral. It is the vibrational matrix elements which introduce the large differences since the geometries of $n\pi^*$ and $\pi\pi^*$ states differ considerably more than the geometries of two $\pi\pi^*$ or two $n\pi^*$ states. When the energy gap is small, of the order of 1 to 2 quanta of the progression-forming vibration, a different if not opposite behavior may be obtained than when the energy gap is large. For the latter situation, vibrational matrix elements relevant to the coupling of $n\pi^*$ and $\pi\pi^*$ states are dominant whereas for small gaps this is no longer necessarily true. However, as pointed out by Hochstrasser [7], when the energy separation of the two interacting electronic states is small the description in terms of Born-Oppenheimer states is no longer a good first approximation to the actual states, and the above argument does not apply.

The polarized absorption spectrum of 1,6-naphthyridine-d_6 in naphthalene at 4 K in the region of the first two excited singlet states is shown in Fig. 3. The lowest singlet system with origin at 28794 cm^{-1} has been assigned $^1A''$ $(n\pi^*)$ [33]. The origin of the second singlet system has not been identified but comparison with the naphthalene spectra and the nature of the 1,6-naphthyridine-d_6 spectrum

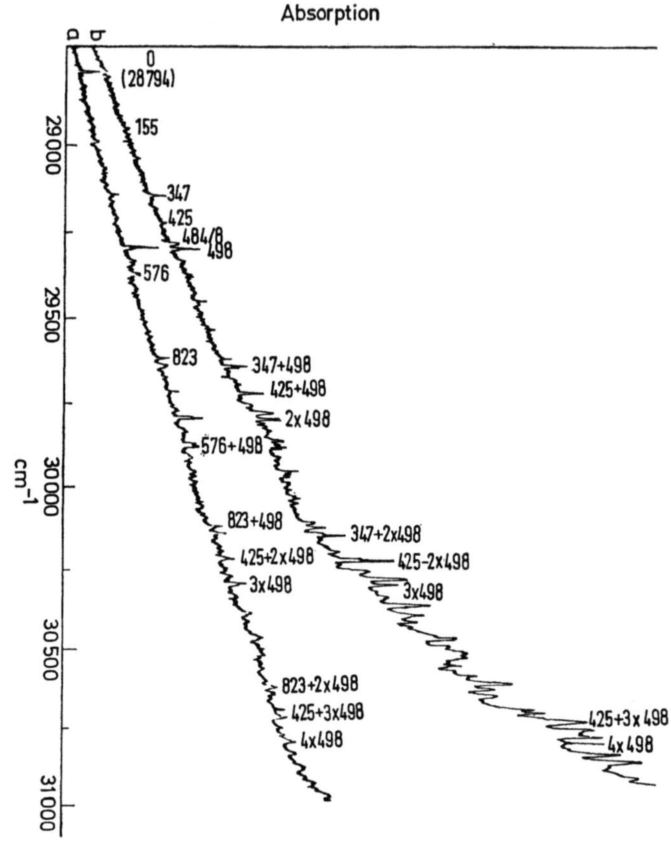

Fig. 3. Polarized absorption spectra of 1,6-naphthyridine-d_6 in naphthalene at 4 K. The spectra encompass the region from the lowest singlet system ($^1n\pi^*$ 0—0 at 28794 cm^{-1}) to the onset of the next singlet system (the lowest $^1\pi\pi^*$)

observed suggests that the first singlet $\pi\pi^*$ state lies within 2000 to 3000 cm^{-1} of the $^1A''$ ($n\pi^*$) state. It can be noted (Fig. 3) that for absorption parallel to the naphthalene b (and c') axes, which predominantly contain the in-plane molecular axes, there is a considerable increase in the number of vibronic bands and in their intensities for vibrational energies more than $\sim$1000 cm^{-1} above the $^1A''$ ($n\pi^*$) origin. A comparable picture is not observed for the a crystal axis, which corresponds mainly to the out-of-plane molecular axis.

In common with other aza-naphthalenes, 1,6-naphthyridine displays a progression in the angle-bending mode. Along the a axis this progression displays a normal Franck-Condon envelope but when based on the false origins 347 and 484/8 polarized parallel to the b and c' axes, this is certainly not the case.

(*b*) *Biacetyl*

Biacetyl is of particular interest because it displays both small- and large- (statistical) molecule behavior over a narrow energy range [34]. This is evidenced by the pressure dependence of its phosphorescence below excitation energies of 22500 cm^{-1}. The transition to large-molecule behavior at this energy has been explained by Drent, van der Werf and Kommandeur [34] as evidence for the presence of another electronic state, most probably the ^3B_g state.

There are four electronic states to be considered in the present discussion. Of these the ^1A_u and ^3A_u states have been located at ~21850 and 20000 cm^{-1}, respectively [35]. In the low-temperature pure-crystal spectra, Sidman and McClure [36] assigned the origins to the bands at 22873 and 20421 cm^{-1}, respectively. The ^1B_g and ^3B_g states, on the other hand, have not been unambiguously observed in spectra. Electric dipole transitions to them from the ground state are forbidden. Calculations [37] have placed the ^1B_g and ^3B_g states at much higher energies than the ^1A_u and ^3A_u states, although from experimental considerations Sidman and McClure [36] believed them to be very close.

The vapor-phase absorption spectrum observed in the region 21800 to 22500 cm^{-1} may be described as a very broad band with superposed poorly contrasting fine structure (<5 cm^{-1}) [34,38]. Byrne and Ross [5] have intimated that biacetyl may be among the largest open-chain compounds to show fine structure. The broadening mechanism is not an intrinsic one but is more probably related to spectral congestion. The appearance of fine structure at regular intervals is evidence of this. The essentially diffuse nature of the spectra must arise principally from the activity of the very-low-frequency torsional mode (~35 cm^{-1}), which is active in both states involved in the transition. The sharpened nature of the spectrum in the low-temperature crystal supports this interpretation, since hot bands and sequence bands no longer contribute to the absorption.

To conclude, small-molecule behavior is predicted [39] for the low-lying vibrational levels of the lowest ^1A_u state, and it is in fact observed. At low pressures irreversible intersystem crossing is not expected because of the relatively low density of triplet vibrational levels in near resonance. The reversible or irreversible nature of the intersystem crossing is not expected to contribute to the line broadening of the vibrational levels of the ^1A_u state, since the observation of fluorescence indicates that the rate of radiationless transition is too small to contribute more to the linewidths than other mechanisms, such as Döppler broadening and environmental effects. At excitation energies greater than 22500 cm^{-1}, line broadening due to electronic relaxation may be envisaged because for these energies biacetyl displays statistical behavior. Preliminary spectroscopic studies are in accord with this prediction since no fine structure of the type observed at energies less than 22500 cm^{-1} is noted [38].

(c) *Benzophenone*

Benzophenone is another example of a molecule showing small-molecule behavior in a specific region of its absorption spectrum [30,31]. Here it concerns intersystem crossing between the lowest excited singlet and triplet states, which are separated by about 2800 cm^{-1}. Very fast intersystem crossing is induced by intermolecular interactions [30]. Under isolated-molecule conditions relative to the radiative lifetime as calculated from the integrated oscillator strength, irreversible behavior is not obtained.

There appears to be general agreement that the lowest excited singlet and triplet states are $n\pi^*$ [40,41]. The lowest $^3\pi\pi^*$ state does not lie between the two $n\pi^*$ states, it is at least 2000 cm^{-1} above the $^1n\pi^*$ state [41]. Dym and Hochstrasser [42] have reported the linewidths of the 0—0 bands of the $^1n\pi^*$ and $^3n\pi^*$ systems as 0.9 and 0.23 cm^{-1}, respectively, in the solid. For the vibronic band $0+1178$ (1 quantum of the —C=O stretching vibration) in the $^1n\pi^*$ system a linewidth of 2.1 cm^{-1} was obtained, corresponding to a relaxation rate of 4×10^{11} sec^{-1}. Therefore the linewidths in the excited singlet system are broadened by intersystem crossing. The energy uncertainty corresponds to a rate of intersystem crossing of $\sim1\times10^{11}$ sec^{-1}. As expected, no fluorescence is observed in the condensed phase.

Rentzepis and Busch [31] have studied the electronic relaxation of benzophenone in benzene solution by picosecond spectroscopy, in particular the intersystem crossing rate at two different energies corresponding to the vibrationless level of the lowest excited state and to a level $v=2$ for the —C=O stretching mode. They explained the variation in the rate of intersystem crossing with excitation energy as resulting from different Franck-Condon factors. On following the disappearance of excited singlet state molecules through singlet–singlet absorption, rates of $\sim2\times10^{11}$ and 5×10^{10} sec^{-1}, respectively, were measured. This is in general agreement with the rate deduced from the linewidth study, although the decreased rate at the higher energy implies a smaller linewidth, certainly an interesting phenomenon.

It is clear that a number of questions need to be answered. Why, in the condensed phase, is the intersystem crossing between two $n\pi^*$ states so efficient? What is the explanation of the conflict between the linewidth studies of Dym and Hochstrasser and the lifetime studies of Rentzepis and Busch, with respect to the vibrationally excited levels? It was in an attempt to provide some answers to these questions that Hochstrasser, Lutz and Scott [43] carried out picosecond experiments on the dynamics of triplet state formation. In benzene solution the build up of the triplet state had a lifetime of 30 ± 5 psec, but this could only be considered as a lower limit of the intersystem crossing rate since vibrational relaxation also contributed to the radiationless transition to the triplet state. The rate of triplet state build-up was found to be solvent-dependent.

Lifetime studies in the gas phase at low pressures indicated two decay rates [31]. The short-lived component ($\tau=10$ μsec) has a lifetime an order of magnitude larger than the fluorescence lifetime determined from the absorption spectrum. The longer-lived component has $\tau\simeq200$ μsec. In the model proposed to account for these results, the short-lived component is said to derive from the strong coupling of the singlet level with only a small number n of triplet levels, such that the observed lifetime is given by

$$\tau = h(\gamma_S/n + \langle\gamma_T\rangle)^{-1} \tag{5}$$

where γ_S and $\langle\gamma_T\rangle$ are the linewidths (average for the triplet) in the isolated molecules for uncoupled levels in each manifold. The coupling leads to a dilution of the singlet character, resulting in an increased lifetime. This behavior is characteristic of the small-molecule model, where anomalously long fluorescence lifetimes are predicted and observed.

In the model considered the strongly coupled levels are themselves weakly coupled to a very large number of levels of the triplet manifold. If this manifold is sufficiently dense, it may provide a dissipative quasi-continuum, so giving rise to a relaxation channel. This is the proposed explanation for the long-lived component. It corresponds to phosphorescence and its yield shows some pressure dependence. However, even at the lowest pressures studied, $\sim 10^{-4}$ torr, the yield of the long-lived component was not negligible, indicating that the triplet manifold does act as a dissipative quasi-continuum. The high-pressure results may be viewed in the same light as the condensed-phase results previously discussed.

Vapor-phase linewidth studies of benzophenone in this interesting spectral region have not been carried out. Nevertheless, it may be possible to make some predictions on the variation of linewidths with pressure. The pressure dependence or collisional deactivation may be qualitatively described by assigning a linewidth $\Delta_{\mathrm{T}}^{\mathrm{V}}$, which is assumed proportional to pressure, to the triplet levels. Thus, the linewidths are given by

$$\gamma = \Gamma_{\mathrm{S}}/n + \Gamma_{\mathrm{T}} + \Delta_{\mathrm{T}}^{\mathrm{V}} \tag{6}$$

where Γ is the radiative width, it being assumed that other relaxation channels, namely electronic relaxation to S_0, the ground state for the S_1 and T_1 states, may be neglected, and that collisions do not appreciably effect the width of the singlet state. Equation (6) shows the role played by the solvent in inducing intersystem crossing. At the one extreme of very low pressures the linewidth is that corresponding to a lifetime of 10 µsec, while at the other extreme–liquid or solid solution and presumably high-pressure vapor — the linewidth is that corresponding to a very short lifetime of 10 psec.

(*d*) *Quinoxaline*

For quinoxaline (1,4 diazanaphthalene) the two lowest excited singlet states are separated by about 5000 cm^{-1}. In both durene and naphthalene crystal [3,44] at 4 K and in the vapor [45,46] at 295 K linewidths as small as 1 cm^{-1} are observed for the first singlet transtion $^1B_1(n\pi^*) \leftarrow {}^1A_1$. In contrast, the second transition ($^1A_1$ $(\pi\pi^*) \leftarrow {}^1A_1$) is marked by broad bands ($\sim$300 cm^{-1}) which upon closer inspection reveal poorly contrasting fine structure [46]. The region to the red of the $\pi\pi^*$ 0—0 band in the *ab* polarized spectrum of quinoxaline-d_6 in naphthalene is shown in Fig. 4. The fine structure consists of a progression of peaks at intervals of $\sim$70 cm^{-1}. This interval is very prominent in the $^1B_1$ spectrum and there it corresponds to the frequency difference between the two totally symmetric progression-forming vibrations, 510 and 580 cm^{-1}. A similar explanation may apply here. The absorption is strongest parallel to the c' and b crystal axes ($I_{c'} > I_b$), which for naphthalene host crystal correspond most often to the molecular long and short axes, respectively, and it is in accord with the in-plane nature of transitions to $\pi\pi^*$ states.

The spectrum observed is not simply a superposition of the sharp $^1B_1$ $(n\pi^*)$ and broad $^1A_1$ $(\pi\pi^*)$ systems, but an interaction spectrum of the two. This is

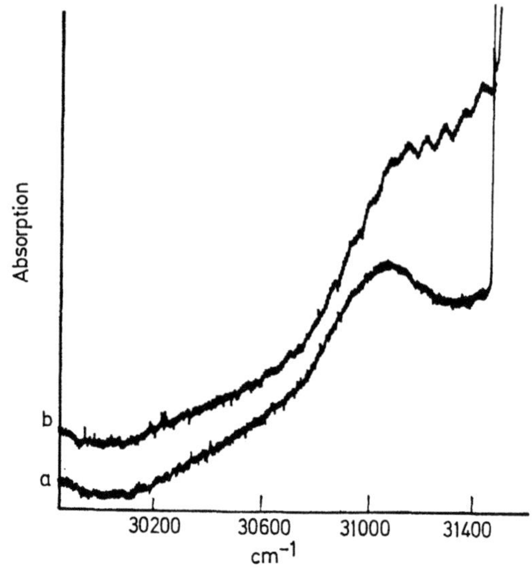

Fig. 4. The *a b* polarized absorption spectra of quinoxaline-d_6 in naphthalene at 4 K. The region towards the onset of the lowest $^1\pi\pi^*$ state is displayed. The lowest singlet state is $^1n\pi^*$ with origin at 25913 cm^{-1}. The sharp cut-off is due to host crystal absorption. Along the *b* axis fine structure is evident and the interval separating the peaks is reminiscent of that observed in the $^1n\pi^*$ system

borne out by its transition polarization, the absorption strength, and the line shapes. Note that in Fig. 4 there is no indication of structural absorption along the *a* or *b* axes towards lower energies. To simulate the spectrum it is necessary to carry out a treatment, yet to be described, in which the interactions between the vibronic states are considered in detail, as has been done for phenanthrene [47].

McDonald and Brus [48] have considered the time-resolved emission in the vapor phase, following excitation in the S_2 ($^1\pi\pi^*$) region (2800–3200 Å). In terms of the lifetime a two-component fluoıescence is recorded: *A* fast component $\tau_F \leqslant 50$ msec, limited by the time dependence of the excitation source, and an intermediate component with a pressure-dependent lifetime in the μsec range. Both appear in the same spectral range. This behavior is reminiscent of that described for benzophenone. However, a major difference concerns the number and nature of the interacting states which, for quinoxaline, are at least three: $S_1 (n\pi^*)$, $S_2 (\pi\pi^*)$ and one or more triplet states.

Apart from the low-resolution vapor spectrum reported by Glass, Robertson and Merritt [46], no other study of the second system is available. It is therefore not possible to discuss line shapes in the vapor-phase spectrum in relation to the decay rates of the two-component fluorescence. More information is available foı the crystal, but in the condensed phase the decay rate of the intermediate pressure-dependent 'fluorescence' may have been considerably increased, as is the case for benzophenone.

(e) Azulene

Any discussion of nonradiative processes and line broadening must at some time make a reference to azulene. It is not our aim to review the wealth of spectroscopy that has been carried out on azulene over the past few years, but rather to describe some recent and very pertinent work by Hochstrasser and Li [49] on linewidths for the $S_1 \leftarrow S_0$ (at 14652) and $S_2 \leftarrow S_0$ (at 28048 cm^{-1}) transitions. The linewidths were measured at 1.2 K for azulene and azulene-d_8 in a naphthalene host crystal.

The well known anomalous fluorescence from S_2 has been interpreted in terms of a much slower radiationless transition out of S_2 than S_1, such that for S_1 the fluorescence lifetime is severely shortened relative to the radiative lifetime. The anomaly is related to the unusual energy disposition of the two lowest excited singlet states. Hochstrasser and Li wished to ascertain whether the spectral linewidths were consistent with this interpretation and also whether the S_1 linewidths of azulene-d_8 were narrowed in comparison, as theoretically predicted. Their results are listed in Table 1. The spectral resolution was claimed to be $\leqslant 0.15$ cm^{-1}; as linewidths in the S_2 system corresponding to the observed fluorescence lifetime are of the order of 10^{-4} cm^{-1}, the linewidths of ~ 0.50 cm^{-1} measured must be considered crystal-imposed. It is assumed that the maximum crystal inhomogeneity contribution to the S_1 linewidth is similarly ~ 0.50 cm^{-1}. This leads to a line broadening due to rapid nonradiative electronic relaxation of 1.61 (-h_8) and 1.27 (-d_8) cm^{-1} as compared to 0.64 cm^{-1} (-h_8) determined by Rentzepis [50] from lifetime studies of azulene in benzene sclution at 300 K.

Table 1. Linewidths for the azulene and azulene-d_8 0—0 bands in the S_1 and S_2 systems in naphthalene host crystal at 1.2 K. Results in cm^{-1}

System	Energy	—h_8	—d_8
S_1	14652	2.08	1.77
S_2	28048	0.47	0.50

(f) Sym-Triazine

The four lowest excited singlet states of s-triazine are predicted to lie within an energy range of ~ 10000 cm^{-1} and to be $n\pi^*$ in character [51]. S-triazine is isoelectronic with benzene but the presence of the three nitrogens gives rise to four low-lying nx^* states in addition to the $\pi\pi^*$ states characteristic of benzene. Only one of these $n\pi^*$ states is associated with an allowed electric dipole transition. The presence of so many electronic states in such a small energy range leads to a very complex situation. This is borne out by the complexity of the observed spectrum, for which only a partial analysis is available [51,52]. An $^1E''$ state is located at 30870 cm^{-1} (crystal 30014) and a $^1A_2''$ state some 5000 cm^{-1} to higher energies. The other $^1E''$ and $^1A_1''$ states have not been identified.

The relatively small number of fundamental vibrations combined with the high molecular symmetry lead to sparse vibrational manifolds for the energy gaps of interest here. Therefore, it would seem that s-triazine should display small-molecule behavior. The spectrum should be characterized, not so much by the broadening of bands, but by the appearance of vibrational structure which cannot be accommodated in a straightforward vibrational analysis based on a single electronic state. This indeed appears to be the case. The vapor spectral analysis is far from complete and a considerable amount of the observed vibrational structure remains unassigned. In the vapor phase, rotational fine structure persists up to 2500 cm^{-1} to higher energy of the 0—0 band [52]. The observation of rotational structure is in accord with the appearance of fluorescence [53].

The pure crystal spectrum at 4 K (Fig. 5) similarly has not been fully analyzed, but it has been noted that the out-of-plane absorption at energies greater than the 0—0 band broadens rapidly while becoming more intense. It has been suggested [52] that in the crystal the line broadening may be a result of vibronic excition–phonon coupling. It is not expected that the crystal spectrum will exactly parallel the vapor spectrum, since the crystal environment affects the disposition of the energy levels and also the molecular configuration. However, it is difficult to reconcile the extensive broadening observed in the crystal with an intramolecular electronic relaxation mechanism, since the broadening is characteristic of the statistical case and the conditions pertaining to the statistical case do not appear to be overtly satisfied.

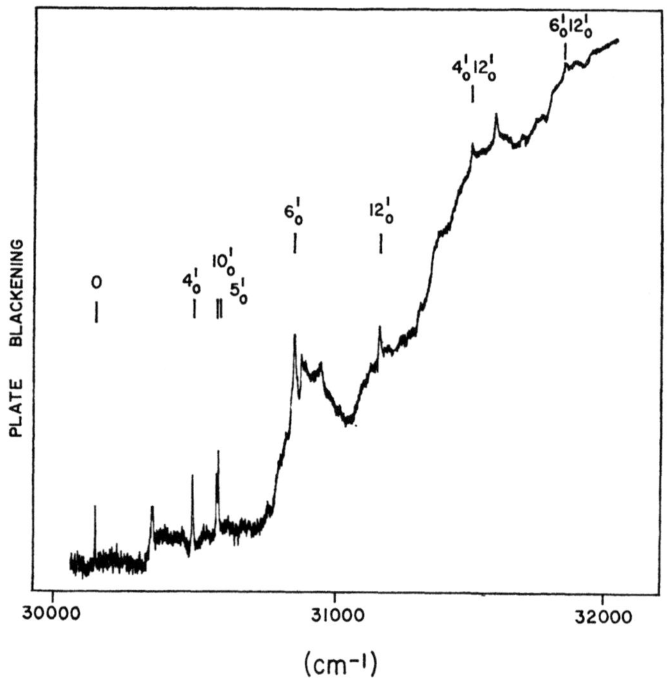

Fig. 5. The pure crystal absorption spectrum of sym-triazine-d_3 at 4 K taken, with the incident light perpendicular to the *ab* cleavage face

(g) *Quinoline and Isoquinoline*

Recently, a number of instances have been reported where the two interacting electronic states are not only close-lying but are nearly degenerate. For quinoline the two lowest excited singlet states of $n\pi^*$ and $\pi\pi^*$ character, respectively, are separated by only 400 cm^{-1} [54]. In the $\pi\pi^*$ system, which is much more intense than the $n\pi^*$ system, the hot band at 0—490 shows reasonably well-developed fine structure, the 0—0 band only poorly contrasting, close-packed fine structure; the bands at higher energy do not display any fine structure whatsoever. A somewhat similar behavior is reported for isoquinoline, whose vapor spectrum is shown in Fig. 6. Linewidths of <1 cm^{-1} are observed for the 0—0 band whereas for energies $\geqslant 800$ cm^{-1} above the origin no apparent fine structure remains. As yet no

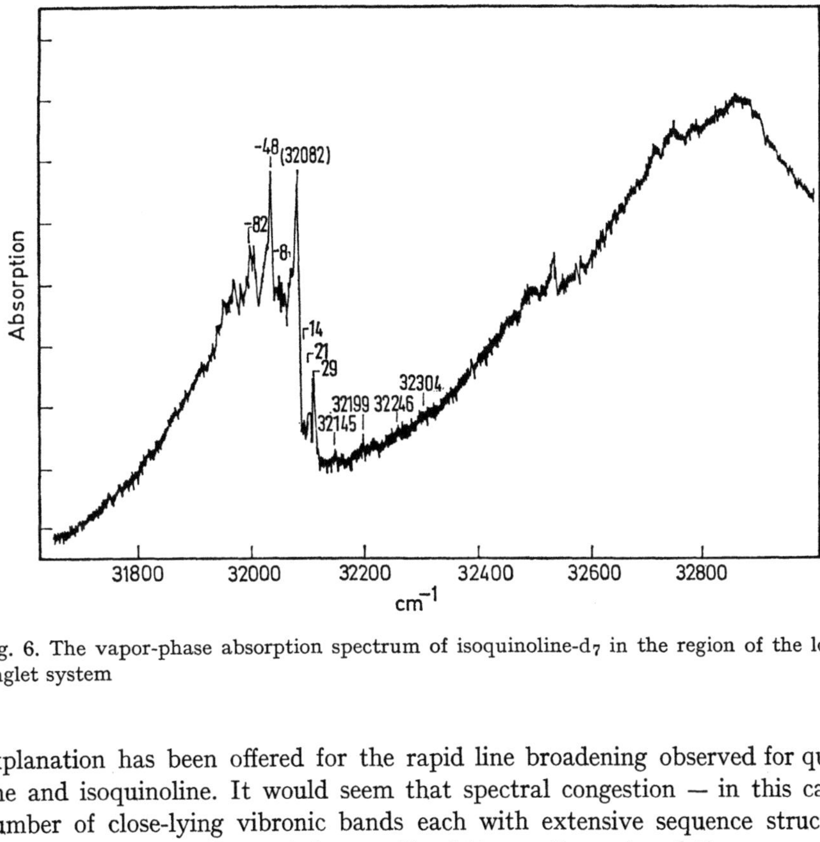

Fig. 6. The vapor-phase absorption spectrum of isoquinoline-d_7 in the region of the lowest singlet system

explanation has been offered for the rapid line broadening observed for quinoline and isoquinoline. It would seem that spectral congestion — in this case a number of close-lying vibronic bands each with extensive sequence structure, combined with the effects of the coupling between the $\pi\pi^*$ and the sparse, relatively low-intensity manifold of the $n\pi^*$ state — may be responsible for the apparent line broadening. It is clear that in the vincinity of the electronic origin small-molecule behavior obtains but for energies greater than $\sim$1000 cm^{-1} above the origin this may no longer be true. There is a gradual transition to large-molecule (statistical) behavior. Hochstrasser [7] would have it that the line broadening is an indication of the presence of another electronic state. This suggestion would be in accord with the observation for quinoline of an $n\pi^*$ state at lower energies than the $\pi\pi^*$ state. The same reasoning suggests that a second electronic state, presumably the $n\pi^*$, is energetically close to the observed $\pi\pi^*$ state of isoquinoline.

This near degeneracy of two low-lying electronic states, particularly the lowest $\pi\pi^*$ and $n\pi^*$ states, is more characteristic of the triplet states. The singlet–triplet spectrum of crystalline acetophenone originates at 26108 cm^{-1} ($^3n\pi^*$) with sharp linewidths (1–3 cm^{-1}). At $0 + 440$ the lines start to become diffuse (>5 cm^{-1}). Dym and Hochstrasser [42] have considered this as evidence for the presence of a second triplet state ($\pi\pi^*$ type) before the onset of broadening, that is below 26550 cm^{-1}. Similarly, for propiophenone in the crystal, line broadening is encountered at about 470 cm^{-1} to higher energies of the sharp origin [42]. However, it is difficult to reconcile the diffuseness with rapid nonradiative transitions, as required by the Dym and Hochstrasser interpretation, since the small electronic energy gap does not comply with the statistical case but rather with the small-molecule description. A better explanation may include coupling with the phonon modes.

For 2,4,5-trimethylbenzaldehyde the lowest triplet state in durene crystals is $\pi\pi^*$, but Migirdicyan [55] has interpreted higher-energy emission observed at 195 K as emission from the $^3n\pi^*$ state 400 cm^{-1} above the $^3\pi\pi^*$ state.

Spectral results have been presented covering the whole range of line broadening in large molecules, from very broad (~300 cm^{-1}) featureless bands to broad but extensively structured, and finally to sharp bands whose widths are governed by radiative and not nonradiative rates. In terms of the nonradiative transitions they correspond to energy gaps between the interacting electronic states ranging from large ~10000 cm^{-1}, to intermediate/small ~4000 cm^{-1}, to near resonance. As the energy gap is increased from ~10000 cm^{-1} to 20000–30000 cm^{-1}, an approximately reverse process occurs but without the appearance of anomalous structure. The smooth, broad, structureless bands are narrowed till eventually the radiative linewidth assumes predominance.

4. Theoretical Models

4.1. Statistical Case

An expression for the line broadening resulting from rapid radiationless transitions has been derived by introducing a number of simplifying assumptions into the physical description of the phenomenon.

a) A Born-Oppenheimer (BO) vibronic state ϕ_s of an higher excited electronic state ψ_2 is imbedded in a manifold of BO vibronic states $\{\phi_l\}$ of a lower excited electronic state ψ_1. A constant interaction is assumed between ϕ_s and each of the manifold states $\{\phi_l\}$:

$$\langle \phi_l | H | \phi_{l'} \rangle = \delta_{ll'} \, E_{l'} \tag{7}$$

$$\langle \phi_l | H | \phi_s \rangle = V \, . \tag{8}$$

b) The manifold $\{\phi_l\}$ is assumed to consist of equally spaced states. The density of states ϱ is given by $\varrho = \varepsilon^{-1}$ where ε is the constant separation between the states.

c) Only the state ϕ_s carries intensity in transitions from the ground state. Transitions to the manifold states $\{\phi_l\}$ are assumed to be negligible because of very small Franck-Condon factors.

d) A further assumption is made, which applies to the statistical case,

$$V \gg \varepsilon .$$

A justification for the validity of the theoretical model proposed to explain rapid nonradiative electronic relaxation is to be found in the agreement between the theoretically derived linewidth and that experimentally observed.

Interaction between the discrete state ϕ_s and the manifold states $\{\phi_l\}$ leads, upon diagonalization of the appropriate Hamiltonian, to vibronic eigenfunctions of the form

$$\psi_n = a_s^n \phi_s + \sum_l b_l^n \phi_l . \tag{9}$$

Since ϕ_s is the only state associated with an oscillator strength the intensity of transitions to the ψ_n are proportional to $|a_s^n|^2$ and the lineshape is given by

$$\varrho |a_s^n|^2 = \frac{\varrho V^2}{(E_n - E_s)^2 + (\pi V^2 \varrho)^2} \tag{10}$$

This is a Lorentzian function with width Δ given by Eq. (1).

It has been of interest to calculate the linewidth theoretically, but this has encountered some problems, particularly in the evaluation of V [56]. In the scheme outlined above, V refers to the nuclear kinetic energy operator $T(Q)$, for which the linear term, Eq. (11), is believed to provide the major contribution

$$V = V_{1l,2s} = \sum_i \int \chi_{1l}(Q) \frac{\left\langle \psi_1(q,Q) \left| \frac{\partial U(q,Q)}{\partial Q_i} \right| \psi_2(q,Q) \right\rangle}{E_2(Q) - E_1(Q)} \frac{\partial}{\partial Q_i} \chi_{2s}(Q)\, dQ \tag{11}$$

where $\phi_s = \psi_2(q,Q)\,\chi_{2s}(Q)$ and $\phi_l = \psi_1(q,Q)\,\chi_{1l}(Q)$ are adiabatic wave functions, $U(q,Q)$ is the potential energy operator of the molecular Hamiltonian, and $E(Q)$ are the adiabatic potential energies. In general, one normal coordinate predominantes in the coupling and hence the summation may be replaced by a single term. In the evaluation of Eq. (11) it has been customary to approximate the electronic part of the integral at the equilibrium configuration Q_0 of the second state ψ_2, and to write the vibrational wavefunctions $\chi(Q)$ as a product of $3N$-6 vibrational wavefunctions, one for each of the normal coordinates. Hence, V may be written as a product

$$V \simeq KS \tag{12}$$

where K and S are given by expressions of the form

$$K_i = \frac{\left\langle \psi_1(q,Q_0) \left| \left(\frac{\partial U(q,Q)}{\partial Q_i} \right)_{Q_0} \right| \psi_2(q,Q_0) \right\rangle}{E_2(Q_0) - E_1(Q_0)} \left\langle \chi_{1l}(Q_i) \left| \frac{\partial}{\partial Q_i} \right| \chi_{2s}(Q_i) \right\rangle \tag{13}$$

and

$$S_i = \prod_{j \neq i} \langle \chi_{1l}(Q_j) | \chi_{2s}(Q_j) \rangle \tag{14}$$

More simply K may be written in the form

$$K \simeq \left[\left\langle \psi_1 \left| \frac{\partial U}{\partial Q_i} \right| \psi_2 \right\rangle \left\langle 1 \left| Q_i \right| 0 \right\rangle \right] \frac{\omega_i}{E_2 - E_1} \tag{15}$$

where the term appearing in the square brackets is familiar from Herzberg-Teller theory and typically has values in the range 100—1000 cm^{-1}. Its magnitude may be estimated from the intensity induced by vibronic coupling [3]. Taking values typical for the aromatic hydrocarbons $\omega_i \simeq 1000$ and $E_2—E_1 \simeq 10000$ cm^{-1} one arrives at $K \simeq 100$ cm^{-1}. Now for energies of 10000 cm^{-1} Sharf and Silbey [56] estimated that $\varrho S^2 \simeq 10^{-5}$ leading to a linewidth

$$\Delta \simeq 1 \text{ cm}^{-1}$$

in obvious disagreement with observation.

In a further work Sharf and Silbey [57] showed that the fault lay not in the theory but in the approximation made in evaluating Eq. (11). Under certain conditions a large error is introduced by approximating the electronic matrix element and the energy difference $E_2(Q)—E_1(Q)$ by their values at the equilibrium configuration of ψ_2.

4.2. Intermediate Case

Naphthalene

Wessel and McClure [16] considered in detail the other extreme, namely the coupling between two close-lying electronic levels. For naphthalene, the two lowest excited singlet states are separated by ~3000 cm^{-1}; hence only a very sparse vibrational manifold of $S_1(^1B_{2u})$ is in near-resonance with the origin of $S_2(^1B_{1u})$. This presents a situation where it is possible, in terms of the size of the matrix to be daignolized, to consider all, or at least an important fraction, of the interacting states lying within a definite energy range about the S_2 origin. This was done by Wessel and McClure. They obtained a calculated spectrum, intensities (peak heights) and energies in good agreement with the observed spectrum.

Their calculations involved diagonalizing a matrix whose input elements were experimentally determined. The diagonal elements recorded the zero-order energies and the off-diagonal elements mesured the coupling between the sparse manifold of S_1 and the discrete levels of S_2. The calculations were refined by varying the input quantities to achieve best fit.

In *p*-xylene, relative to durene host crystal the naphthalene S_1—S_2 energy gap is reduced by 500 cm^{-1} to ~2000 cm^{-1}. The crystal spectra at 2.2 K are shown in Fig. 7. Additions of a_g vibrations to the $^1B_{2u}$ electronic state are unperturbed by the $^1B_{1u}$ system even where there is initial overlap of the two systems. Bandwidths are in the range 0.5 (0—0) to 2.0 cm^{-1}, increasing to higher wavenumbers. In contrast, lines of the $^1B_{2u}$ system but with overall B_{1u} symmetry (short axis polarized) intensify and broaden as they approach resonance with the S_2 origin.

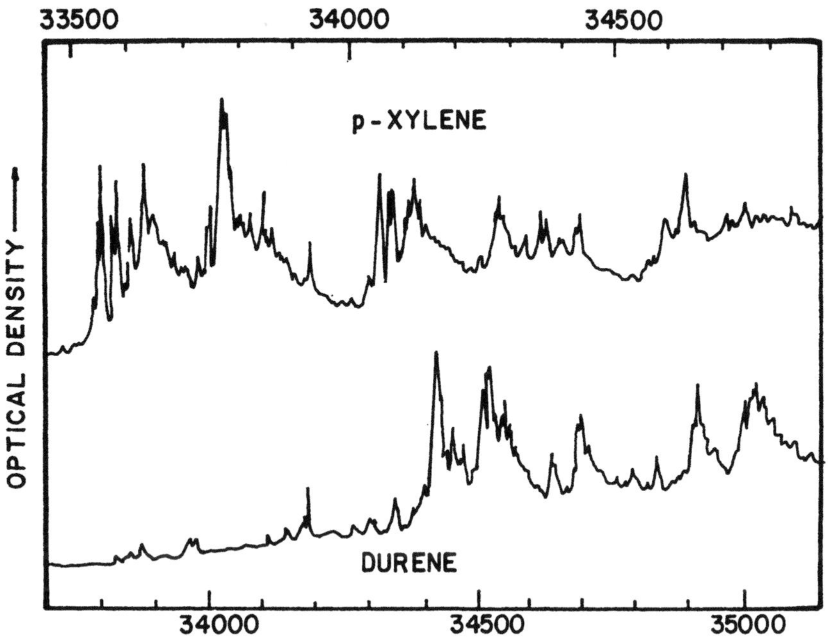

Fig. 7. The absorption spectra of naphthalene in *p*-xylene and durene host crystals at 2.2 K, in the region of the 0—0 band of the second singlet system. As shown, the origins of the first singlet system are coincident. The results are from Wessel and McClure [16)]

The vibrational manifold of S_1 at energies of ~2000 cm^{-1} above the S_1 origin is in near resonance with the origin of the S_2 system in a *p*-xylene host crystal but ~500 cm^{-1} off resonance in a durene host. The naphthalene spectrum in the latter host provides the off-resonance (zero-order) spectrum from which may be calculated the interaction spectrum observed in *p*-xylene host. Of course, it must be recognized that the intensities and energies of the levels appearing in the off-resonance spectrum predominantly record the coupling with a number of vibronic levels of S_2 and not just the origin, and hence it is necessary to arrive at estimates of the individual couplings since it is these that are required as input quantities in the matrix problem.

Of the eight b_{3g} fundamentals that may be active in the first-order coupling two, the C—H stretching modes are immediately eliminated because their frequencies are greater than the energy gap. Of the remaining six, two are experimentally observed. In setting up the interaction matrix, not all the interacting states are included, but only those active in the off-resonance spectrum. This considerably reduces the scope of the matrix problem. There is also a restriction placed on the number of vibronic levels, a_g additions to $^1B_{1u}$, of the $^1B_{1u}$ state considered.

These estimates are necessarily approximate since not all the interacting levels of the S_2 electronic state nor of other electronic states are included. By allowing for variation in the parameters making up the zero-order spectrum (that is before the interaction is switched on) some compensation is made for this shortcoming.

It has been shown that, for the two extreme situations, namely very large and very small energy gaps between the two interacting electronic states, models have been proposed which successfully predict the interaction spectra. In the intermediate range the problem is more difficult to handle. Here it is not acceptable to consider an average coupling and an average density of states on the one hand, nor on the other hand is it feasible to carry out an analysis in which all the interacting states are included since their number is far too great. The absorption spectra of the second excited singlet states of quinoxaline (Fig. 4) and phenanthrene (Fig. 8) clearly illustrate the case. Essentially broad bands are obtained but they are characterized by poorly contrasting fine structure.

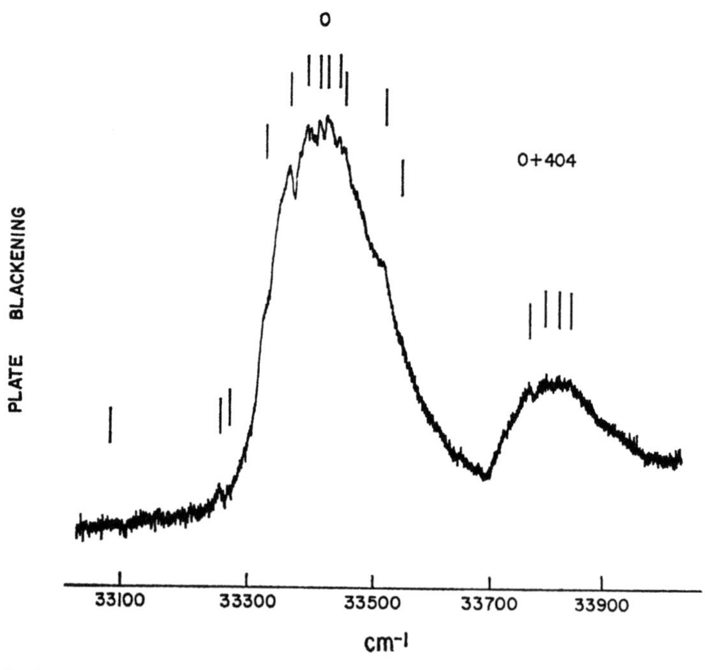

Fig. 8. The 0—0 and 0 + 404 (a_1) bands in the b polarized $^1B_2 \leftarrow {}^1A_1$, 3000 Å) absorption spectrum of phenanthrene in durene at 4 K. The appearance of discrete structure is indicated

4.3. Phenanthrene

A treatment of the $S_2(^1B_2 \leftarrow {}^1A_1)$ absorption spectrum of phenanthrene has recently been presented [47]. There are too many interacting states to permit carrying out a full treatment in which every state is accounted for uniquely. On the other hand, the statistical approach of assigning a constant coupling V and a constant separation ε washes out the details of fine structure. In the approach adopted, the dense manifold of vibrational states of S_1 in near resonance with S_2 is partitioned into a sparse manifold which contains a small number of discrete levels, those associated with the largest coupling, and the remaining dense manifold. First the interaction between S_2 and the small number of discrete levels of S_1 is considered. This is followed by a consideration of the broadening of these levels through interaction with the remaining high density of states of S_1.

An interesting consequence of this approach concerns the absorption intensities of the discrete but broadened lines. The broadened lines all appear with the same heights (maximum extinction coefficients) but differ in widths. However, when the intensities of all the lines are summed, this aspect is obscured.

In the scheme where one level, S_2, associated with an oscillator strength interacts with a sparse manifold $\{l\}$, the intensity carried by each of the resulting discrete states ψ_n is proportional to the squared coefficient $|a_s^n|^2$, where a_s^n is defined by Eq. (9). These discrete states are broadened by interaction with the remaining dense manifold and a Lorentzian line shape $L_n(E)$ of the form Eq. (10) is assumed for each broadened line. Note that these lines are still associated with the same intensities $|a_s^n|^2$, that is

$$|a_s^n|^2 = \int_{-\infty}^{\infty} L_n(E)\, dE\,. \tag{16}$$

Since the coupling between ψ_n and the remaining dense manifold depends on the coefficients a_s^n, the interaction V appearing in Eq. (10) is different for each ψ_n but related to a_s^n. Therefore the lineshapes of the final broadened lines are given by

$$L_n(E) = (a_s^n)^2 \frac{\varrho(Ca_s^n)^2}{(E_n - E)^2 + (\pi\varrho(Ca_s^n)^2)^2} \tag{17}$$

where C is some constant.

The sparse vibrational manifold of S_1 in near resonance with the S_2 vibrationless level has been constructed on the basis of the observed $S_1 \leftarrow S_0$ and $S_2 \leftarrow S_0$ absorption spectra. The intensities, Franck-Condon factors and vibrational frequencies in the two systems are required. As in the Wessel-McClure treatment for naphthalene, an interaction matrix is set up but in a different manner, with the diagonal elements representing the zero-order energies of the discrete state s and the sparse manifold states $\{l\}$ and the off-diagonal elements representing the coupling between them. Diagonalization of the matrix leads to the new sparse manifold of discrete states ψ_n. By computing the effective density of states corresponding to the S_2—S_1 energy gap (4500 cm^{-1}) and assigning a magnitude to the constant C in Eq. (17), one obtains a calculated spectrum (Fig. 9). C governs the strength of the coupling responsible for the statistical broadening; it is a measure of the mean value of all the vibronic couplings and a parameter that can be varied to achieve best fit with experiment. Likewise, the energy of S_2 is varied to achieve best fit, since it cannot be directly measured in experiment.

Overall agreement is good between the observed (Fig. 8) and calculated (Fig. 9) spectrum for the region near the vibrationless level of the second excited singlet state $^1B_2$ of phenanthrene, although not all the fine structure is faithfully

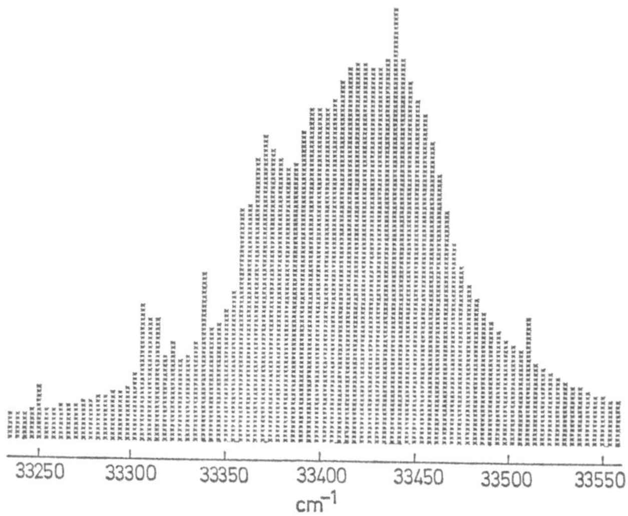

Fig. 9. The computed absorption contour of the 0—0 band of the $^1B_2 \leftarrow {}^1A_1$, 3000 Å system of phenanthrene

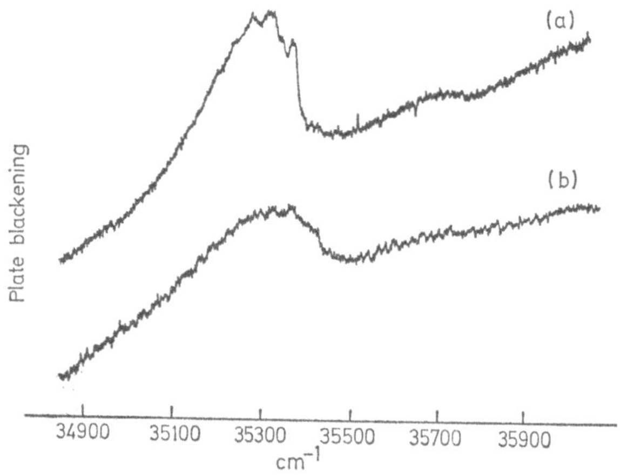

Fig. 10. The vapor-phase absorption spectra of (a) phenanthrene and (b) phenanthrene-d_{10}, showing the structure recorded for the 0—0 bands of the 3000 Å system

reproduced. The spectrum depicted in Fig. 8 is for phenanthrene in durene crystals at 4 K.

The phenanthrene vapor-phase spectrum has also been recorded (Fig. 10).[47] Because the S_2—S_1 energy gap is increased to 6000 cm^{-1} and because sequence bands are active, a different fine-structured spectrum is obtained. On the high-energy side of the 0—0 band maximum the observed structure results from the S_1 and S_2 near-resonance interactions and is reinforced by coincident sequence bands. On the long wavelength side the sequence bands fall over a range of frequencies and the overlapping of all these bands, each with its own, but similar, interference structure, leads to a smearing out and consequent loss of fine structure. The effect is more marked for phenanthrene-d_{10} (Fig. 10), possibly because of greater sequence-band activity resulting from smaller vibrational frequencies. The greater energy gap relative to the durene host crystal also leads to a reduction in the prominence of the fine structure.

4.4. Energy Dependence of Density-Interaction Product

Both Langhoff and Robinson [58] and Delory and Tric [59] have considered the effects on the emission decay curves and on the absorption lineshapes of variations with energy in the density of states and in the coupling strengths of the dense manifold with the intensity-carrying discrete state. The former workers also extended their treatment to include coupling with a second manifold, the radiation field continuum.

Langhoff and Robinson have expressed the absorption coefficient $A'(E)$ in terms of the energy-dependent density-of-states function $D_0(E)$ and the energy-dependent interaction $V(E)$:

$$\bar{A}'(E) = \frac{8\pi}{3c} E \mu_{01}^2 \frac{\pi|V(E)|^2 D_0(E)}{(E - \bar{E}_1)^2 + [\pi|V(E)|^2 D_0(E)]} . \tag{18}$$

Instead of assuming that the product $|V(E)|^2 D_0(E)$ is a constant, as is normally done, the energy dependence of the density-interaction product was written in terms of a modifying function $f'(E)$:

$$|V(E)|^2 D_0(E) = |V(\bar{E}_1)|^2 D_0(\bar{E}_1) f'(E) \tag{19}$$

where $\bar{E}_1$ is the adjusted energy of the discrete state. Finally, when expressed as a reduced absorption coefficient $\bar{A}$, the following expression is obtained:

$$\bar{A} = \frac{f(\xi)}{\xi^2 + f^2(\xi)} \tag{20}$$

where

$$\xi = \frac{E - \bar{E}_1}{\pi V(\bar{E}_1)^2 D_0(\bar{E}_1)} \qquad \text{and} \qquad f(\xi) = f'(E) ,$$

which gives the lineshape in terms of the modifying function $f(\xi)$. A similar procedure leads to an expression for the case when coupling with the radiation field is included. They proceded to consider the perturbations of the line shape and decay curves due to the different energy dependences of (a) the interacting continuum and (b) the manifold of discrete states.

For a constant interaction continuum a pure Lorentzian function is obtained for the lineshape. When the continuum is allowed to vary exponentially the Lorentzian becomes asymmetric. For the coupling of the primary state (associated with an oscillator strength) with a manifold of discrete states and with the radiation continuum, a structured lineshape is obtained. The radiation field continuum may be replaced by any other interacting continuum, and in this sense the model approaches that discussed in the phenanthrene lineshape problem. However, the model lacks complete generality in that the manifold of discrete states is not fully random.

4.5. Fano Lines

In contrast to the general rule that transitions to the higher-lying states are severely broadened by intramolecular electronic relaxation, Rydberg transitions in the gas phase appear surprisingly narrow [60,61]. For example, for benzene in the gas phase, the $3R$ Rydberg levels have linewidths of only a few cm^{-1} whereas the linewidths of the $^1B_{1u}$ and $^1E_{1u}$ systems are two orders of magnitude greater. Similar observations have been made for naphthalene [62] and anthracene [61]. The explanation most commonly proposed is that the radiationless transitions in the isolated molecule are inefficient, and this is rationalized on the grounds that there is only small overlap between the large-radius Rydberg orbital and the ground-state π orbital.

An aspect of particular interest in the case of Rydberg levels is the high probability that transitions to them overlap $\pi\pi^*$ systems which carry oscillator strength in the region of overlap. If this occurs, then the model in which only the discrete state carries oscillator strength while the interacting dense manifold does not, no longer applies to the description of line broadening. Because both the discrete state and the interacting manifold carry oscillator strength, interesting interference effects may arise [63]. This problem has been considered by Jortner and Morris [63] on the basis of the Fano [64] formalism for the appearance of antiresonances under the conditions of autoionization, and experimental results have been reported which show some agreement with their conclusions [62].

Jortner and Morris [63], Sharf [65], and Rice and co-workers [66,67] all employ a three-state model in which a discrete Rydberg state is quasi-degenerate with a sparse, intensity-carrying manifold of vibrational levels of a close-lying $\pi\pi^*$ electronic state, overlapping a dense manifold of a lower-lying $\pi\pi^*$ state. Jortner and Morris have considered (1) the broadening of the sparse manifold through interaction with the underlying near-continuum and (2) the interaction between the discrete Rydberg state and the broadened sparse manifold.

In the case of one discrete state interacting with a manifold, the lineshape is given by

$$L(\varepsilon) = (q + \varepsilon)^2/(1 + \varepsilon^2) \qquad (21)$$

where ε is the energy parameter in units of the formal linewidth $\Delta = 2\pi V^2 \varrho$ and q is the line-profile index

$$q = \frac{\vec{\mu}_{ts}}{\pi \varrho V \vec{\mu}_{tl}} \tag{22}$$

It is assumed that the manifold consists of equally spaced levels with energy separation ϱ^{-1}, and that the interaction V of the discrete state s with the manifold states l is constant. $\vec{\mu}_{ts}$ and $\vec{\mu}_{tl}$ are the transition moments to the discrete and manifold states, respectively, it being further assumed that $\vec{\mu}_{tl}$ is a constant for the manifold. As $\vec{\mu}_{tl} \to 0$, the lineshape index q tends to infinity, that is, to a symmetric Lorentzian form. On the other hand, when the discrete state carries no intensity, that is $\vec{\mu}_{ts} = 0$ and therefore $q = 0$, a depression, or symmetric antiresonance, centered at the energy of the discrete state is obtained in the absorption band.

It is clear that interference effects covering the whole range of behavior of q should be of common occurrence. Yet only a few examples of antiresonances have been documented and, of these, only those for the Rydberg spectrum of naphthalene have been assigned with any real confidence [66]. Furthermore, for none of the assigned antiresonances is there a complete cancellation of intensity at the trough of the lineshape. Florida, Scheps and Rice [67] have attempted to provide an explanation for these observations. They suggested that in the three-state model the order of diagonalization should be reversed. First, the interaction between the sparse manifold and the discrete state should be considered and then the interaction between the resulting discrete states and the near continuum. However, it was pointed out that this scheme lacks complete generality since it only provides an explanation for incomplete cancellation of intensity for half the cases [14,68].

Nitzan and Jortner [69], and more recently Nitzan [68] have shown that, if coupling with the radiation field continuum is included, then an incomplete cancellation of intensity may obtain. If this is done, the following expression is obtained for the lineshape [68]:

$$L(\varepsilon) = \frac{(\varepsilon + qY^{\mathrm{L}})^2 + (q^2 Y^{\mathrm{L}} + 1)(1 - Y^{\mathrm{L}})}{(\varepsilon^2 + 1)} \tag{23}$$

where Y^{L} is the quantum yield for the nonradiative decay of the discrete state into the near continuum. When the nonradiative decay of the discrete state is much faster than its radiative decay — a condition very characteristic of internal conversion — $Y^{\mathrm{L}} \simeq 1$. However, as pointed out by Nitzan [68], this analysis is not really relevant to the problem since generally $Y^{\mathrm{L}} \simeq 1$ and then Eq. (23) reduces to the lineshape expression (21) of Fano.

Sharf and Fischer [14] have developed a new approach to account for the lack of a complete cancellation of intensity at the dip in the Fano line, in terms of the Fano treatment of a discrete state and a number of continua. It was shown that the Bixon-Jortner description of the quasi-continuum as a manifold of equally spaced levels with constant interactions with the discrete state and with constant transition moments from the ground state, does not comply with the quantum-mechanical sum rules characteristic of the physical situation. In a real manifold, as distinct from the Fano continuum,

(i) the levels are not all equally spaced,

(ii) the interactions are not all equal, $V_{sl} \neq V'_{sl}$,

(iii) the transition moments $\vec{\mu}_{il}$ are not all of equal magnitude,

(iv) the interactions cannot all be of equal sign while simultaneously requiring that the respective transition moments are all parallel and in the same sense.

In the Sharf and Fischer treatment the real manifold is subdivided into n idealized submanifolds where each one conforms with the Bixon-Jortner model. By taking suitable linear combinations of the n idealized submanifolds the problem may be reduced to that of a discrete state and two manifolds, both of which carry intensity but only one of which is associated with a non-zero interaction. The Bixon-Jortner procedure applied to this situation leads to a Fano lineshape superposed on a constant background absorption.

In this case three parameters are employed to describe the lineshape:

$$q = \frac{\vec{\mu}_s}{\pi \varrho \sum_J \vec{\mu}^J V^J} \tag{24a}$$

$$\Delta = 2\pi\varrho \sum_J (V^J)^2 \tag{24b}$$

$$R_{\mathrm{I}} = (\sum_J \vec{\mu}^J V^J)^2 / \sum_J (\vec{\mu}^J)^2 \sum_J (V^J)^2 \tag{24c}$$

where R_{I} is the fraction of the total manifold intensity associated with the interacting manifold. The summations are over the n idealized submanifolds and it is these summations, satisfying the quantum-mechanical sum rules, which provide the values of the interaction squared V^2, the transition moment squared $\vec{\mu}^2$, and the product $V\vec{\mu}$. It is in the product term $V\vec{\mu}$, that differences arise with the Bixon-Jortner procedure, and hence in the description of Fano lines. Although

$$\sum_J (\vec{\mu}^J)^2 = \vec{\mu}^2 \tag{25a}$$

and

$$\sum_J (V^J)^2 = V^2 \, , \tag{25b}$$

it is not generally true that

$$\sum_J \vec{\mu}^J V^J = \vec{\mu} V \tag{26}$$

where $\vec{\mu}$ and V are constant and average transition moments and interactions for all the manifold states. Clearly, when the manifold carries no intensity, that is $\vec{\mu}^J = 0$, both treatments lead to the same result for the lineshape, namely a Lorentzian of width

$$\Delta = 2\pi\varrho\sum_J (V^J)^2$$
$$= 2\pi\varrho V^2 .$$

A similar approach has been described for the problem of two discrete states interacting with a quasi-continuum, all of which carry intensity [14]. However, in this case six quantum-mechanical rules of the type of Eq. (25) must be obeyed and the problem cannot be simply reduced to that of two discrete states and one idealized manifold, as has generally been done. Even when no oscillator strength is associated with the quasi-continuum, it is still not possible to use the two discrete states and one idealized manifold description.

5. References

1) McClure, D. S.: J. Chem. Phys. *22*, 1668 (1954).
2) Craig, D. P., Fischer, G.: Trans. Faraday Soc. *63*, 530 (1967).
3) Hochstrasser, R. M., Marzzacco, C.: J. Chem. Phys. *49*, 971 (1968).
4) Porter, G., Wright, F. J.: Trans. Faraday Soc. *51*, 1205 (1955).
5) Byrne, J. P., Ross, I. G.: Australian J. Chem. *24*, 1107 (1971).
6) Byrne, J. P., Ross, I. G.: Can. J. Chem. *43*, 3253 (1965).
7) Hochstrasser, R. M.: Accounts Chem. Res. *1*, 266 (1968).
8) Jortner, J., Rice, S. A., Hochstrasser, R. M.: Advan. Photochem. *7*, 149 (1969).
9) Callomon, J. H., Parkin, J. E., Lopez-Delgodo, R.: Chem. Phys. Letters *13*, 125 (1972).
10) Small, G. J.: J. Chem. Phys. *52*, 656 (1970).
11) Neporent, B. S.: Opt. Spectry. (USSR) (English Transl.) *32*, 133 (1972).
12) Bixon, M., Jortner, J.: J. Chem. Phys. *48*, 715 (1968).
13) Haarhoff, P. C.: Mol. Phys. *7*, 101 (1963).
14) Sharf, B., Fischer, G.: Chem. Phys. *3*, 397 (1974).
15) Bixon, M., Jortner, J.: J. Chem. Phys. *50*, 3284 (1969).
16) Wessel, J., McClure, D. S.: Abstracts Fifth Molecular Crystal Symposium, and J. Wessel, Ph. D. thesis, University of Chicago (1970).
17) Wannier, P., Rentzepis, P. M., Jortner, J.: Chem. Phys. Letters *10*, 193 (1971).
18) Wannier, P., Rentzepis, P. M., Jortner, J.: Chem. Phys. Letters *10*, 102 (1971).
19) Dunn, T. M.: Studies in chemical structure and reactivity, Chapter 6. London: Methuen and Co. 1966.
20) Broude, V. L.: Opt. Spectry. (USSR) (English Transl.) *30*, 46 (1971).
21) Pantos, E., Hamilton, T. D. S.: Chem. Phys. Letters *17*, 588 (1972).
22) Brith, M., Lubart, R., Steinberger, I. T.: J. Chem. Phys. *54*, 5104 (1971).
23) Katz, B., Brith, M., Sharf, B., Jortner, J.: J. Chem. Phys. *52*, 88 (1970).
24) Burland, D. M., Castro, G., Robinson, G. W.: J. Chem. Phys. *52*, 4100 (1970).
25) Burland, D. M., Castro, G.: J. Chem. Phys. *50*, 4107 (1969).
26) Colson, S. D., Bernstein, E. R.: J. Chem. Phys. *43*, 2661 (1965).
27) Shimakura, N., Fujimura, Y., Nakajima, T.: Bull. Chem. Soc. Japan *46*, 1297 (1973).
28) Bree, A., Vilkos, V. V. B.: Spectrochim. Acta *27A*, 2333 (1971).
29) Nitzan, A., Jortner, J., Kommandeur, J., Drent, E.: Chem. Phys. Letters *9*, 273 (1971).
30) Busch, G. E., Rentzepis, P. M., Jortner, J.: J. Chem. Phys. *56*, 361 (1972).
31) Rentzepis, P. M., Busch, G. E.: Mol. Photochem. *4*, 353 (1972).
32) Craig, D. P., Hollas, J. M., Redies, M. F., Wait, S. C., Jr.: Phil. Trans. Roy. Soc. London *253A*, 543 (1961).
33) Fischer, G.: J. Mol. Spectry. *49*, 201 (1974).
34) Drent, E., van der Werf, R. P., Kommandeur, J.: J. Chem. Phys. *59*, 2061 (1973).
35) Drent, E., Kommandeur, J.: Chem. Phys. Letters *14*, 321 (1971).
36) Sidman, J. W., McClure, D. S.: J. Am. Chem. Soc. *77*, 6461 (1955).
37) Swenson, J. R., Hoffman, R.: Helv. Chim. Acta *53*, 2331 (1970).
38) Fischer, G.: unpublished results.
39) Parmenter, C. S., Poland, H. M.: J. Chem. Phys. *51*, 1551 (1969).
40) Dym, S., Hochstrasser, R. M., Schafer, M.: J. Chem. Phys. *48*, 646 (1968).
41) Batley, M., Kearns, D. R.: Chem. Phys. Letters *2*, 423 (1968).
42) Dym, S., Hochstrasser, R. M.: J. Chem. Phys. *51*, 2458 (1969).
43) Hochstrasser, R. M., Lutz, H., Scott, G. W.: Chem, Phys. Letters *24*, 162 (1974).
44) Jordan, A. D., Fischer, G., Rokos, K., Ross, I. G.: J. Mol. Spectry. *45*, 173 (1973).
45) Fischer, G., Jordan, A. D., Ross, I. G.: J. Mol. Spectry. *40*, 397 (1971).
46) Glass, R. W., Robertson, L. C., Merritt, J. A.: J. Chem. Phys. *53*, 3857 (1970).
47) Fischer, G.: Chem. Phys. *4*, 62 (1974).
48) McDonald, J. R., Brus, L. E.: Chem. Phys. Letters *23*, 87 (1973).
49) Hochstrasser, R. M., Li, T.: J. Mol. Spectry. *41*, 297 (1972).
50) Rentzepis, P. M.: Chem. Phys. Letters *3*, 717 (1969).
51) Brinen, J. S., Goodman, L.: J. Chem. Phys. *31*, 482 (1959).
52) Fischer, G., Small, G. J.: J. Chem. Phys. *56*, 5934 (1972).

53) Knight, A., Parmenter, C. S.: unpublished results.
54) Youker, J.: Dissertation Abstr. Intern. *30*, 5458 (1970).
55) Migirdicyan, E.: Chem. Phys. Letters *12*, 473 (1972).
56) Sharf, B., Silbey, R.: Chem. Phys. Letters *4*, 423 (1969).
57) Sharf, B., Silbey, R.: Chem. Phys. Letters *4*, 561 (1970).
58) Langhoff, C. A., Robinson, G. W.: Mol. Phys. *26*, 249 (1973).
59) Delory, J. M., Tric, C.: Chem. Phys. *3*, 54 (1974).
60) Wilkinson, P. G.: Can. J. Phys. *34*, 596 (1956).
61) Angus, J. G., Morris, G. C.: J. Mol. Spectry. *21*, 310 (1966).
62) Angus, J. G., Christ, B. J., Morris, G. C.: Australian J. Chem. *21*, 2153 (1968).
63) Jortner, J., Morris, G. C.: J. Chem. Phys. *51*, 3689 (1969).
64) Fano, U.: Phys. Rev. *124*, 1866 (1961).
65) Sharf, B.: Chem. Phys. Letters *5*, 456 (1970) and *5*, 459 (1970).
66) Scheps, R., Florida, D., Rice, S. A.: J. Chem. Phys. *56*, 295 (1972).
67) Florida, D., Scheps, R., Rice, S. A.: Chem. Phys. Letters *15*, 490 (1972).
68) Nitzan, A.: Mol. Phys. *27*, 65 (1974).
69) Nitzan, A., Jortner, J.: J. Chem. Phys. *57*, 2870 (1972).

Received September 23, 1974

Author Index Volumes 26—66

The volume numbers are printed in italics

GPSR Compliance

The European Union's (EU) General Product Safety Regulation (GPSR) is a set of rules that requires consumer products to be safe and our obligations to ensure this.

If you have any concerns about our products, you can contact us on ProductSafety@springernature.com

In case Publisher is established outside the EU, the EU authorized representative is:

Springer Nature Customer Service Center GmbH
Europaplatz 3
69115 Heidelberg, Germany

Batch number: 10620881

Printed by Printforce, the Netherlands